Sexuality in Plants and Its Hormonal Regulation

M.Kh. Chailakhyan
V.N. Khrianin

Sexuality in Plants and Its Hormonal Regulation

Edited by Kenneth V. Thimann
Translated by Vanya Loroch

With 47 Illustrations

Springer-Verlag
New York Berlin Heidelberg
London Paris Tokyo

M.Kh. Chailakhyan
V.N. Khrianin
Plant Physiology Institute
Academy of Sciences of the U.S.S.R.
Moscow, U.S.S.R.

Kenneth V. Thimann (*Editor*)
Thimann Laboratories
University of California
Santa Cruz, California
U.S.A.

Vanya Loroch (*Translator*)
Division of Natural Sciences
Thimann Laboratories
University of California
Santa Cruz, California
U.S.A.

Sexuality in Plants and Its Hormonal Regulation by M.Kh. Chailakhyan and V.N. Khrianin was originally published in Russian by Nauka, Moscow, U.S.S.R., 1982; translated with permission.

Library of Congress Cataloging-in-Publication Data
Chaĭlakhyan, M. Kh.
Sexuality in plants and its hormonal regulation.
Translation of: Pol rasteniĭ i ego gormonal′naia reguliatsiia.
Includes index.
1. Plant sex hormones. 2. Plants, Sex in.
I. Khrianin, V. N. (Viktor Nikolaevich) II. Title.
[DNLM: 1. Plants. 2. Sex. 3. Sex Hormones.
QK 827 C434p]
QK827.C4613 1986 581.1′66 86-33907

Typeset by Publishers Service, Bozeman, Montana.
Printed and bound by Halliday Lithograph, West Hanover, Massachusetts.
Printed in the United States of America.

9 8 7 6 5 4 3 2 1

ISBN 0-387-96488-6 Springer-Verlag New York Berlin Heidelberg
ISBN 3-540-96488-6 Springer-Verlag Berlin Heidelberg New York

Foreword

This is an unusual book. It deals with and clarifies an important but somewhat neglected field of plant science. Our knowledge of the control of sexuality in plants is in complete contrast to that of sexuality in animals, which has been a subject of active and highly fruitful research for decades. In plants, indeed, it is a field that had its roots in the remote past, but for some reason has attracted little attention from modern researchers, although it bears closely on the agricultural and horticultural sciences and other practical applications.

There is a carved stone monument from the 9th Century B.C., which is thought to show the fertilization of the date palm by shaking male flowers over the tree. Although the process that went on was probably not at all understood, this was the earliest realization that within a species, some plants produce flowers of one kind and some give flowers of another kind. In scientifically modern times it was Rudolf Camerarius, working in the botanical garden at Tübingen, who found that in plants bearing two types of flowers, the formation of seeds in one type of flower depended on the simultaneous presence of the second type. The result was about the same whether he used corn (a monocot) or castor bean (a dicot). He observed that the second type of flower produced a "powder" that is essential for seed initiation, and thus recognized the essential nature of pollination and, correspondingly, that the style and ovary could be considered as female organs. Following his publication, *De Sexu Plantarum Epistola* (1694), numerous other examples of male and female flowers were brought to light. The independence of the two plant types was made clear by an experiment of Johann Gleditsch, Director of the Botanic Garden of the Berlin Academy, who in 1764 fertilized a female palm growing alone in that garden with pollen from a male tree growing in Leipzig.

It was soon recognized that there are three types of plants in regard to sexuality: those in which male and female flowers are borne on separate plants ("dioecious"), those that bear male and female flowers on the same plants ("monoecious"), and those in which the male and female organs are borne together in the same flower ("hermaphrodite"). There are a few quantitative variations, as in certain species of the maple tree and some hazelnuts, of which some individuals can be predominantly, but not wholly, of one sex – and there are also cases (e.g., corn) where a few female flowers may be formed among an otherwise fully male cluster.

But it is one thing to recognize the existence of these different types of flowers, and quite another to investigate the causes of such differentiation. It is the special service of the authors of this book that they have devoted a number of years, firstly to bring together all the partial indications and possible causes that have been brought forward over the years, some more convincing than others, and secondly to devise and develop specific experiments to test these numerous indications. As a result, they have been able to build a clear-cut picture of the several hormonal influences that together control sex expression. But not content with that, they have gone an important step further in that they have managed to duplicate the effects ascribed to specific substances by removing the organs that form these substances and replacing them with pure products. This makes their conclusions firm.

As a result, we have before us now, at least in outline, a well-integrated presentation of the chemical influences that determine whether flower initials shall become male or female. The controls appear to be basically the same irrespective of family or genus, monocot or dicot, long-day or short-day plants, although day-length can be a sex-controlling factor of its own. The resulting highly satisfying generalization represents a major contribution to the whole field of plant reproduction.

One remaining aspect to which attention seems not yet to have been directed is that of the bisexual or hermaphroditic flower. Should we interpret the data presented here as indicating that instead of an excess of one or another hormone, it is a delicate balance between them that results in a bisexual flower? This would seem the logical conclusion from the clear results in the latter half of the book; however, our logic is not always Nature's. The widespread occurrence of hermaphroditic flowers suggests that the required balance is not so delicate, but is rather precisely controlled, perhaps by some special stabilizing factor yet to be discovered. At any rate, it is difficult to believe, especially in the face of the wide variation in hormonal ratios indicated by the present work on specifically male and female flowers, that the exact balance needed for hermaphroditism would come about so frequently, without some special controlling influence. Perhaps there is something like a spring lock, which automatically falls into place even though the forces acting on it are unevenly balanced.

There is another respect in which the work described in this book is notable. Indeed, in some sense, it is almost revolutionary. This is in regard to what is indicated as the functions of the hormones. In other work on hormonal function, we have tended to think, and to adduce evidence, at the cellular level: stimulation of enzymes in the cell wall or the middle lamella, reorientation of microtubules, changes in the permeability of the plasmalemma or other membranes, or activation of cell division. In the control of sexuality, we see the same hormones acting as whole organ determinants. Formation of male or female flowers must involve complex contrasts in differentiation. Thus the functions of the plant hormones in the processes described here seem unusual, and we discern an even broader picture than before of the extent of hormone interaction in the whole plant.

Lastly, the book introduces Western readers to a mass of work published in Russian and unavailable till now in a Western language. The sheer volume of work in this field that has been done in the U.S.S.R. may be a surprise to many. I have added a few Western references, but have done it sparingly because I preferred to let the emphasis on the Russian work stand out.

I must take this opportunity to commend the admirable work of Vanya Loroch, who not only made the original translation, but accepted patiently my many changes, improvements, and last-minute afterthoughts that were intended to make the text more readable. A further chore was the necessity of Anglicizing the extensive Russian bibliography. We hope that this translation, by making the remarkable work of Chailakhyan and Khrianin available to Western readers, will form a valuable addition to our knowledge of the flowering process and will add the recognition of still another role of the phytohormones in the control of plant development.

Kenneth V. Thimann
University of California, Santa Cruz

Preface

The continuing progress in fundamental studies of the effect of phytohormones on growth and development in plants has opened up the possibility of directly addressing the problem of the role of these hormones in the regulation of the expression of sex. Many authors have made contributions to this problem with different types of experimental investigations of the different aspects of sexuality and its expression in plants.

That phytohormones may play a role in plant sex regulation first appeared possible almost 50 years ago, after it was demonstrated that hormonal substances inducing the formation of flowers appeared in the leaves of plants exposed to appropriate day lengths (Chailakhyan 1937). Because flowering is closely controlled by day length in a very large number of plants, including particularly the monoecious short-day plants with either hermaphroditic (e.g., Chrysanthemum and Perilla) or unisexual (e.g., corn and cucumber) flowers and also in dioecious plants (e.g., hemp and spinach), it seemed possible that the hormonal substances controlling flowering could indeed act as the sexual hormones of plants. This concept was also based on an analogy with animal physiology, in which the study of sexual hormones had by that time already gained general recognition.

However, this suggestion depended upon later evidence that soon led to the concept of a florigen complex that conditions flowering. With the discovery of gibberellins—phytohormones that (among other actions) induce the flowering of a number of annual and perennial plants—the idea of a florigen complex was divided into two groups of substances: gibberellins, influencing the formation of flowering stems and the mitotic activity of the apex, and anthesins, acting directly on the formation of the flowers themselves (Chailakhyan 1958).

Recently, many authors have shown that all the known phytohormones (auxins, gibberellins, cytokinins, abscisic acid, and ethylene) influence to varying extents the expression of sexuality in plants. The majority of those experiments tested both the immediate and delayed effects of applied phytohormonal preparations by spraying the plants' surface organs with solutions at various concentrations. Because a number of contradictory conclusions resulted from these experiments, it became necessary to build a logical system.

We thus felt that the problem of hormonal regulation of sexuality in plants needed to be approached once more on the basis of the rules that had been developed while simultaneously investigating the hormonal regulation of flowering itself. Accordingly, three main questions were addressed: 1) When (i.e., at what age) is the sex expressed, 2) Where (i.e., in which organs) is it expressed, and 3) What are the accompanying changes in the metabolism of hormones? This book contains an account of these three approaches, embodying the results of experimental and theoretical work on the regulation of sexuality in plants.

In the first three chapters, we discuss the evolution and the genetics of sexuality, including sexual dimorphism. We analyze the literature and our own data pertaining to the effect of primary environmental factors on sex expression in plants, and we present the results of experiments on the influence of phytohormones on the expression of sexuality as a function of age in plants. In Chapters 4 to 6, we describe first an integral model of the expression of sex that clarifies both the role of individual organs and of the substances they synthesize, and their application to sex determination both in dioecious plants and in those monoecious plants that bear unisexual flowers. Later, we present further results of analytical work on the content of biologically active phytohormones in plant development and in differential sex expression. Last, we consider the general concept of genetic and hormonal regulation of the expression of sex in plants.

M.Kh. Chailakhyan
V.N. Khrianin

Contents

Introduction

The problem of sex in plants has been addressed since the times of Empedocles (circa 485–455 B.C.), Aristotle (384–322 B.C.), and Aristotle's student Theophrastus (370–322 B.C.). In their philosophical treatises, they were the first to note the presence of sex in plants, making a direct link between the flower and the fructification, and drawing an analogy between the reproduction of plants and the reproduction of animals. Camerarius (1694) discovered sexual differentiation in plants, Linnaeus (1729, 1753, 1767; cited by Bobrov 1970) created the concept of sexual organization in plants, and Kölreuter (1761) demonstrated empirically the existence of sex in plants through his classic experiments on true hybrids. The discovery of double fertilization in angiosperms by the Russian scientist Navashin (1898) was received with great interest by the scientific community.

The formation and modification of sex in plants, however, have been relatively little studied, and yet these phenomena have considerable theoretical and practical significance. We have only to realize the possibilities that could emerge in selection work and in agriculture if the processes of regulation of sex expression in our crop plants could be mastered. Vavilov (1932, p. 19) wrote on this subject: "In order to develop a method of selection, the study of sex in plants and the establishment of the variety of sexual types have decisive importance; this subject should be addressed for all the important cultivated plants."

The investigation of sex, as pointed out by Sabinin (1963), is an important sector in the general area of the physiology of development of a plant organism. Sabinin and his co-workers Minina, Gusiova, and Satarova studied the effect of mineral nutrients and atmospheric composition on the modification of sexual characters in plants (Sabinin 1940; Minina 1952). The studies of Molotkovskyi (1957, 1960, 1968, 1976) were important in elucidating the regularities in the sex differentiation of plant tissues. The evolution of sex as well as the physiological and biochemical aspects of sex were studied in great detail by Djaparidze (1963, 1965; English translation 1967, 1969*).

*L. I. Djapardize (1967), Sex in Plants, Part 1, 197 pp.; (1969), Sex in Plants, Part 2, 214 pp., Israeli Program for Scientific Translations, Jerusalem.

In the last 30 years, phytohormones have been studied not only for their effects on growth and development (Rakitin 1950; Chailakhyan 1958; Kulayeva 1973; Muromtsev and Agnistikova 1973; Kefeli 1973; Gamburg 1976), but also for their influence on thc manifestation of sex in plants (Heslop-Harrison 1957; Frankel and Galun 1977; Sidorskyi 1978; Minina and Larionova 1979).

The basis for this book is the several years of experimental research that we have devoted to the elucidation of regulatory mechanisms in plant sexuality. Complex investigations on whole plants, the culture of isolated embryos, and a developed integral model of sex expression have helped to answer three basic questions:

1. When do the generative primordia on the apex of the plants studied undergo differentiation?
2. What is the role of other organs in the initiation of male and female flowers on the plant?
3. What hormones are synthesized in association with the expression of sex, and at what level of realization of the genetic information is this expression accomplished?

The study of hormonal regulation of sex expression both in dioecious plants and in monoecious plants with unisexual flowers, and an understanding of the interactions between phytohormones and the genetic apparatus, can provide a basis for the development of entirely new methods of directing plant development and the yield of the harvest.

1
Sex in Plants and Factors in Sexual Differentiation

A. The Evolution of Sex

The evolution of the plant and animal worlds led to the evolution of the means of reproduction. Sexual reproduction became the most progressive form of reproduction in the organic world. The significance of the sexual process in phylogenesis resides in the fact that fertilization leads to an organism with dual heredity, which in turn ensures greater stability and adaptability to the constantly changing conditions of life. Herskowitz (1965) pointed out that even if reproduction were only asexual, the earth would still be populated by genetically distinct organisms. Each variant would arise as a result of a mutation in the ontogenesis of the previous individual, which in turn would have arisen from an uninterrupted line of generations. However, such a direct inheritance is of limited effectiveness, because it relies on the occurrence of rare mutations. Compared with asexual reproduction, the development of sexual processes offered immense genetic advantages, as obligatory genetic recombination greatly accelerated the rate of evolution.

The development of the sexual process doubtless occurred in several steps during the course of evolution. As early as 60 years ago, Goldschmidt (1927) and later, Vendrovskyi (1933), came to the conclusion that living forms of lower organization are monoclinous, and that dicliny appeared phylogenetically after hermaphroditism. They were able to show that during the evolution from monoclinous to diclinous* forms, several stages of variable hermaphroditism occurred in which hermaphroditic forms acquired sometimes female and sometimes male sexual organs during their development. The subsequent stage of evolution brought rudimentary hermaphroditism in which one of the sexes gradually disappeared and diclinous forms appeared. Goldschmidt (1927) thought that in the case of monoclinous forms, there was no antagonism between the genes of the opposite sexes, that the antagonism appeared in hermaphroditic forms that had

---

*Diclinous = having stamens and pistil in separate flowers; dioecious = having male and female flowers on separate plants [Ed.].

bisexual tissues, and that in rudimentary hermaphroditism even elements of the opposite sex disappeared and strictly unisexual forms appeared. The opinion of Grishko (1935) was that the antagonism between the sex-determining genes is caused by their various interactions after a mutation. Such an evolutionary pathway from hermaphroditic to diclinous plants has been supported by numerous Soviet scientists (Vavilov 1920; Rozanova 1935; Grossgeim 1945; Krechetovitch 1952; Minina 1952; Takhtadjian 1964; Zhukovskyi 1964, 1967; Djaparidze 1963, 1965/English translation 1967, 1969).

This concept of the pattern of evolution is supported by the presence of numerous intermediary forms, ranging from hermaphroditic to diclinous species. There are reports in the literature on the existence of unisexual forms of the castor bean (Sidorov and Sokolov 1945, 1947) and the sunflower (Kuptsov 1935). The evolutionary passage from hermaphroditism to dicliny is demonstrated by the formation of monoclinous flowers or intersexes on monoecious plants (Kardo-Sisoeva 1924; Grishko 1935; Sukachev 1938; Minina 1965). Further evidence arises from the indications that bisexual are more primitive than unisexual flowers (Hutchinson 1926–1934, 1959; Heinz 1927). The interconversion of stamens and pistils (Masters 1862; Penzig 1921; Vuillemin 1926) proves the common origin of the structures of the androecium and gynoecium in individual development and it further helps to elucidate some fundamental questions of phylogeny (Takhtadjian 1941; Kozo-Polyanskyi 1937, 1950). Thus, the manifestation of sex in plants is the result of a balance or lack of balance between the growth of the androecium and the gynoecium; this is expressed by primordia that (in many cases) appear at a very early stage of development (Heslop-Harrison 1963).

Djaparidze (1965) wrote that in the present era, a number of hermaphroditic plants having either bisexual flowers or both male and female flowers on the same plant are already physiologically diclinous. Surikov (1965) suggested that the evolutionary origin of dicliny both in plants and animals could be explained by a reorganization of the self incompatibility system; during the process of evolution, hermaphroditic forms came to experience incompatibility between their sexual elements after fertilization, and this eventually led to the appearance of dicliny. Using biochemical methods in the study of evolution of sex in plants, Sidorskyi and Sidorskaya (1974) concluded that the formation of dioecious plants from hermaphroditic forms began with biochemical differentiation of plant populations. The opinion of Timofeev-Resovskyi et al. (1977) is that the appearance of dicliny and therefore of sexual reproduction is the major evolutionary acquisition after the appearance of DNA reduplication.

There is no doubt that diclinous plant forms have a whole array of advantages over hermaphroditic forms, including in particular a higher range of adaptability to environmental conditions (Minina 1952), some of the biological properties of seeds (Soroka and Zhatov 1970; Sidorskyi 1971), overall productivity (Gogina 1971), viability of the sexual elements (Safaryan 1966; Fursa, 1969), longevity of the vegetative period (Dobrunov 1935a,b; Lvova 1959; Mizunov 1968), and numerous physiological and biochemical indices (Djaparidze 1965; Sidorskyi et al. 1971a,b; Sidorskyi and Sidorskaya 1974).

The problems of the evolution of sex in plants are complex, and the path of the sexual evolutionary process as described may not be the only one. For instance Khokhlov (1946, 1949) suggests that in the future development of plants, the normal sexual process will be replaced by apomixis. Although some authors (e.g., Rozanova 1948) share this opinion, the majority of researchers (Komarov 1940; Yuzepchuk 1958; Djaparidze 1965) think that apomictic species are to be regarded as being in the course of regression or even disappearing because apomictic reproduction has a restrictive effect on formative and adaptive processes in plants (Minina 1966). There is in nature a multitude of pathways and forms of evolution of sex, and the evolutionary approach to the problem of sex allows a deeper understanding and appreciation of the role of those factors that determine sex expression.

B. The Genetic Theory of Sex Determination

Modern genetics originated in the first years of the 20th century (Mendel 1866; (Russian translation, Mendel 1965). De Vries in Holland (1889, 1901, 1903), Correns in Germany (1900a,b, 1902a,b, 1907), and Tschermak in Austria (1900, 1904, 1914) produced results showing that the laws of Mendel were not limited to the garden pea, but applied to many other plants as well. The genetic basis for the determination of sex was first demonstrated by Correns (1906) in experiments in which male and female plants of dioecious bryony (*Bryonia dioica*) were crossed between themselves and with monoecious bryony (*Bryonia alba*). Further research in this direction led to the discovery of sex chromosomes, or heterochromosomes, designated as X or Y chromosomes (Morgan 1924, 1928, 1937). The existence of such a chromosomal mechanism allows regulation of the number of male and female individuals in many species.

Genes responsible for sex determination are not located solely on the sex chromosomes but on autosomes as well (Muller 1922, 1927; Morgan 1934). The sex determining factor is the balance between the influences of all these genes. Thus, one might say that in the XX-XY type, the X chromosome contains more female genes, while the autosomes contain more male genes. The Y chromosome, however, is indifferent. Although change in sex can be promoted by the action of a single pair of genes, usually this change is a result of interaction between several or many pairs of genes. Thus, sex in plants is a polygenic character.

Variation of the ratio of genes that participate in sex determination is a function of variation in the chromosomal set. Providing that the balance of genes remains unchanged, addition or loss of an entire set of chromosomes does not result in change of sex. Changes in the numbers of particular chromosomes that lead to intermediate values of the balance of genes, however, lead to the formation of intermediate sexual types—intersexes. Changes leading to a ratio that exceeds the norm result in the appearance of supersexes (Herskowitz 1968). This theory is not merely a supposition based on the presence of two X chromosomes in females and one X chromosome in males, but relies on experimental evidence showing that normal ratios are disturbed when unusual chromosomal complexes appear, as in triploids, intersexes, supersexes, etc. (Morgan, 1937). After the creation by Morgan (1928, 1937) of the chromosomal theory, sex determination became linked, first with the genotype and second with the influence of environmental factors. When the sex is determined by the genotype, it is often possible to correlate the differences between male and female individuals with the differences in chromosomal sets. In the second type of control, genes still play an important role as they allow the cells to react to environmental conditions in a specific way (Herskowitz 1968).

Differences between sex chromosomes have been found in a number of plants, mostly in dioecious forms. However, these differences appear to be less striking than in animals. Heteromorphous X and Y chromosomes were described in the liverwort *Sphaerocarpus donnellii* by Allen (1932; cited by Lvova 1963), in campion (*Melandrium album*) by Blackburn and Winge (1923), and in Elodea (*Elodea canadensis*) by Santos (1921–1924). The process of sexual determination was examined in detail in such typical dioecious plants as hemp (*Cannabis sativa*) and spinach (*Spinacia oloracea*). Initially imperfect methods in the experiments of Strasburger (1910) and McPhee (1924) prevented the discovery of heterochromosomes in these plants. Only during later cytogenetic investigations were discernible heteromorphous X and Y chromosomes established in hemp (Hirata 1927; Breslavets 1933; Driga 1934; Hoffmann 1941, 1952; Yamada 1943; Köhler 1964a,b).

The inheritance of sex in hemp and the mechanism of sex determination in this plant offer a rather complex picture (Westergaard 1958). Hemp possesses 2n=20 chromosomes. All nuclei of vegetative cells in male individuals contain 18 autosomes plus X and Y chromosomes, while female individuals have 18 + 2X. The male gametes are heterochromosomal, therefore each pollen grain contains in its nucleus, in addition to 9 autosomes, a sexual X or Y chromosome (9 + X or 9 + Y); the grains carrying X are female sex determinants and those carrying Y the male ones. Female plants, on the other hand, are homochromosomal and the oösphere nucleus always contains 9 autosomes plus the X chromosome which carries female sex determinants; in other words, normal male and female plants under natural conditions carry XY and XX chromosomes, respectively, and the ratios of male to female individuals are, as a rule, equal.

In addition to this, however, hemp contains allelic Xm genes, which decrease the induction of femaleness. Plants with XXm chromosomes produce inflorescences of female type but are not necessarily female. Indeed, depending on the presence of additional genetic and nongenetic modifiers, female hemp plants can become male. Plants having XmXm still retain female inflorescences but function more like males. Based on data obtained with polyploid hybrids, Köhler (1964) came to the conclusion that genes enhancing maleness are autosomal and are counterbalanced by the X chromosomal genes that enhance femaleness, the Y chromosome being indifferent with respect to sex-regulating genes. Such a variety of types can account for the great lability of plant form and the high variability of sexual characters in hemp. Change in autosomal maleness as well as in X-chromosomal femaleness can be brought about by methods of selection and also appear spontaneously by mutations.

Sex determination in spinach (containing 2n=12 chromosomes) is controlled by a series of alleles:- Y, Xm, and X (Rosa 1925; Thompson 1955; Dressler 1958, 1973; Janick and Iizuka 1962; Akkos 1965). The combination XX leads to female plants, XmX or XmXm to intersexes, and XY or XmY to male plants. Genetic factors cause further modifications that result in the formation of intersexes from XY and even YY type plants.

In certain species, the chromosomal differences related to sex appear to be significantly more complex than in hemp or spinach. In some cases the distinctions can affect not merely one chromosomal pair but two; in others, the heterochromosomes are represented by complex fragments. Thus in male *Rumex* plants the homologue of the X chromosome consists of two fragments: Y1 and Y2. In two species of *Humulus* plants of one sex, each carries two pairs of heteromorphous chromosomes, whereas in plants of the other sex, all the pairs are identical (Lvova 1963).

Heterochromosomes are found in more than 45 species of angiosperms (Rozanova 1935), but are not found in 26 species of dioecious plants. Sometimes heteromorphous chromosomes are found in hermaphroditic plants. However, when *Melandrium* plants are hermaphroditic, their chromosomal set contains one heteromorphous chromosomal pair, and Belar's opinion (1926) is that such plants arose as a result of sex modifications of male plants. Having analyzed similar cases in a number of plants, Boysen-Jensen (1938) came to the conclusion that sexual chromosomes are a result, and not the cause, of sex differentiation. Lvova (1963) believes that the absence of heterochromosomes is not by itself an indication that sexual forms are absent or that their qualitative differences are minimal. The plants could still have a tendency towards the development of a given sex, even though the qualitative differences are not expressed in the structure of the chromosomes.

The sex differentiation scheme described above (XX-XY types) does not provide a complete explanation for the formation of bisexual flowers nor for the formation of flowers of opposite sex in dioecious plants. In this respect, Goldschmidt's theory (1927) of the determination and inheritance of sex is of interest. According to this theory, male as well as female individuals contain in their common haploid set of autosomes (n) a male sex gene or a complex of genes, M. On the other hand a more potent female sex gene, F, is located on each of the two X chromosomes. Additionally there is still another female sex gene, f, that is less potent than M. At times the Y chromosome includes the gene, f, and at others it lacks the sex-determining gene. The dominance relations between these genes are: FF > MM > Ff. According to Goldschmidt's theory, in dioecious plants the female individuals have the composition (2n)XX = MMFF, and the male individuals have (2n)XY = MMFf. Consequently, each contains in its genotype an influence making for the opposite sex. Thus, Goldschmidt's genetic balance theory of genotypic sex determination (1915) states that the sex is determined by a quantitative ratio or balance of factors in sexual cells that characterize male or female sex. Primary and secondary sexual characters are established through the balance between the two competing genetic states, and because they are quantitative, in nature the balance can be modified in either direction.

A different and relatively peculiar view of the nature of sexual distinctions in higher plants was offered by Correns (1928a,b). In his opinion, bisexual higher plants possess a gene, A, determining the development of staminate flowers; a gene, G, determining the development of pistillate flowers, and a gene (or genes), Z, determining the location and time of development of genes A and G. During the transition from bisexualism to dicliny, this gene complex was supplemented (probably as a result of mutations) with two additional genes: realizer *a* for the A complex and realizer *y* for the G complex. According to Correns, these realizer genes *a* and *y* decrease the expression of genes of the opposite sex: *a* decreases the effectiveness of G, and *y* similarly decreases the activity of A. The functioning of these realizer genes is considered to be dependent upon environmental factors, as the location and time of sex determination may vary greatly (Wettstein 1924a,b; 1937).

If we can consider it established that one of the sexes is heterozygous while the other is homozygous, we can see that this supplies a reason for the generally observed equal ratio between the sexes. This equal ratio in many dioecious plants provides the basis for a genetic explanation of sex determination, namely as a potentially bisexual trend in the development of sexual cells, organs, or individuals towards the male or female type (Riger and Michaelis 1958). This trend can be influenced by external factors, but there is no doubt that sex determination is genetically controlled (Frankel and Galun 1977; Chailakhyan and Khrianin 1980). Sex determination is followed by the formation of male or female cells or organs during the embryonic stage and later development of the individual, i.e., by sex expression. It is this expression that is subject to wide fluctuations during plant development—fluctuations that are caused by environmental conditions in the natural situation or by artificial changes induced experimentally. Sex reversal or sex transformation, i.e., a complete change from one sex to the other, can occur because of natural, pathological, or preset factors (Riger and Michaelis 1967). Moreover, as noted above, cases exist where an individual possesses a chromosomal formula characteristic of one sex but phenotypically expresses the other.

The hypotheses and theories described were engendered by the complexity of the genetic control of sex differentiation in plants. At the same time, they show that sex differentiation in plants is not determined *solely* by the XX and XY chromosomes. Initially it was shown in *Melandrium* (Westergaard 1958) that the formation of stamens depends upon specific genes in the Y chromosome of male plants, which are absent in female plants. However, it was established later (Frankel and Galun 1977) that the genetic information for stamen formation is also present in female *Melandrium* plants. The role of the Y chromosome is believed to be exclusively regulatory (Mittwoch 1969, 1973). Indeed, it is unlikely that the formation of staminate and pistillate flowers in dioecious plants can be related solely to the transfer of information contained in XX or XY chromosomes. Numerous cases have been reported where acceleration or delay of plant development results in the formation of flowers of the opposite sex or of intersexes on initially typical male or female hemp and spinach plants (McPhee

1924; Schaffner 1928; Valter and Lilienstern 1934; Grishko 1935; Khrianin and Chailakhyan 1977).[1]

The preceding experiments and concepts show that while the *determination* of sex is based upon the genetic apparatus, sex *expression* is dependent not only upon the genotype, but also upon environmental conditions and internal metabolic changes.

C. Sexual Dimorphism

Sexual dimorphism in different plant groups has been the subject of numerous studies (Manoylov 1924; Joyet-Lavergne 1931; Ivanov 1935; Grishko 1935; Valter et al. 1940; Naugolnykh 1945; Minina 1962; Lvova 1963; Djaparidze 1965; Minina and Larionova 1979). These studies encompassed morphological, biochemical, and physiological[2] aspects of differences between male and female plants. The general conclusion from all these researches is that sexual dimorphism cannot be regarded as a mere adaptive character developed in the process of evolution. On the contrary, it is also a consequence of physiological and biochemical differences between male and female organisms. Secondary sexual distinctions are coded by different genotypes in the opposite sexes, and are genetically linked to the primary sexual characters.

The discovery of heterochromosomes in certain plants provides a morphological marker for the differences between the gametes, which in turn determine subsequent metabolism and sex expression. In dioecious plants, sexual dimorphism manifests itself not only through differences at the morphological level but also through anatomical, biochemical, and physiological properties of individuals of different sex. Male and female plants differ from one another in the duration of growth and development as well as in other respects. However, while sexual distinctions in invertebrate animals are clearly expressed, some dioecious plants (wild strawberry, cloudberry, and others) display virtually no morphological distinctions except flower structure, although cases of extreme sexual dimorphism do exist (hemp, spinach, poplar, hop, asparagus, white mulberry).

The identification of the sex at early stages of development in plants such as hemp, spinach, sea buckthorn, and others would be not only of theoretical interest but also very important for plant production. However, despite numerous

[1]Tissue culturing can have similar effects. A striking instance was reported with cultures of a variety of birch (*Betula pendula*) that forms male flowers from the terminal bud and female flowers from lateral buds. Yet when these lateral shoots were cultured, the resulting plants produced only male flowers (Huhtinen and Yahyaoglu, Silvae Genetica 23:1–3, 1976) [Ed.].

[2]Much of this early work relating to metabolism as a controlling factor is reviewed extensively in the two volumes by L. I. Djaparidze: Sex in Plants (in English), 1963 and 1969 (National Science Foundation, Washington, D.C.) [Ed.].

attempts, first by Sprecher (1913) and later by Morozov (1920), Satyanarayana (1934), and Grishko (1935), it has not been possible to find distinctions between the morphological characters of male and female dioecious plants at early stages of growth. Distinctions in general structure and in individual morphological characters become apparent only at flowering time when the sex of plants can be easily determined by their habit. In spinach, for example, female plants (unlike males) grow to relatively large size, bear a large number of leaves, and have an extended life span (Lvova 1963). In another dioecious plant, hemp, morphological distinctions are even better expressed, and have been well studied (Makarevich 1935a,b; Chailakhyan 1937; Senchenko et al. 1963). Male hemp plants are more elongated, with a relatively thin stem, long internodes, and leaf blades gradually decreasing in size toward the top of the plant where they are reduced to fairly small linear leaves. Both the volume and the surface of the root system of male plants are about one third the size of those of females. During the period of flower formation, female plants have a more powerful and compact stem with close, short internodes and bigger leaves with blades much wider than those of male plants. Male and female plants start flowering almost at the same time, but the male plants finish blossoming and die off after 10 to 20 days. Female plants continue to vegetate, increase in height and in stem thickness, accumulate substances necessary for seed formation, and complete their growth some 30 to 50 days after the males.

Besides morphological differences, a series of distinctive features has been established in the anatomy of the stems in male and female hemp (Makarevich 1935; Senchenko et al. 1963). Stems of male plants have a less compact arrangement of phloem bundles and individual fibers, a larger cavity of phloem fibers, and a poorly lignified xylem. The phloem fiber tissues remain underdeveloped because of the shortened vegetative period. Elementary fiber cells have variable shapes and inner cavities. Stems of female plants are more lignified, and have well-developed primary and secondary fibers as a result of intense activity of the pericycle and the cambium, which continues throughout their longer vegetative period. Rapid growth of female elementary fiber cells during the second half of the vegetative period in the female plant results in cells whose walls are well filled out with cellulose, having regular shapes and slitlike cavities. All this ensures high strength and quality of hemp fiber in the female. Clear secondary sexual characters have also been described by Darwin (1939) in many species of *Restionaceae*.[1]

Morphological and anatomical distinctions apply not only to herbaceous but also to many woody plants such as the tung tree (*Aleurites*), willow (*Salix*), poplar (*Populus*), and others. In particular, male specimens of black poplar (*Populus nigra*) are bigger than females of comparable age, their branching is more extensive, and the buds are bigger; however, the leaves on female trees are considerably longer than those on male trees (Starova 1969).

[1]A small family related to the rushes (*Juncaceae*), mainly limited to South Africa and Australasia; they have much reduced flowers, minimal leaf surface and rough fibrous stems [Ed.].

Male and female plants differ in a whole array of biochemical indices characterizing basic metabolism (Ivanov 1935). For instance, it has been shown that the protein content of male hemp leaves is lower than that of female leaves, and that male plant proteins contain more arginine and lysine, whereas female plant proteins contain more histidine and tyrosine (Kizel, Pashkevich 1937).

The experiments of Kubarev (1966) showed that the nucleic acid content (DNA and RNA) is considerably higher in the flower clusters and leaves of female hemp and spinach plants than in those of males. However, it was established later (Slonov 1974a) that this ratio changes during the process of development of both male and female plants.

Differences in enzymes also have been reported. Thus, male and female plants differ with respect to peroxidase (Tadocoro 1930, 1933; Djaparidze, Kezely 1934; Kezely 1942, 1944; Marutyan 1954; Ostapenko 1960; Penel 1976), catalase (Marutyan 1954; Ryazanskaya 1956; Lushinskyi 1963; Djaparidze 1965), oxidative activities (Minenkov 1924; Djaparidze, Kezely 1934), and other enzymes. It was found also that enzymatic activity in male and female individuals is not a constant but varies according to the species, age, and the stage of growth.[1]

Numerous studies have sought to determine whether there is an interrelation between sexual dimorphism and content of pigments, in particular of carotenoids. In their work on Mucoraceous fungi Schopfer (1928) and Köhler (1934) showed that female mycelia turn yellow due to their high content of carotene, whereas male mycelia remain white. Analyses of several organs of dioecious higher plants (hemp, spinach, and sea buckthorn) revealed also that the content of carotenoids tends to be higher in female individuals than in males (Lebedev 1948).

Sex differentiation involves not only carotene, but other carotenoids as well (Kun 1941); thus the leaves of female hemp contain more total carotenoids, including carotene, lutein, and violaxanthin, than leaves of male plants (Slonov 1974b).

Occasionally, individual male and female plants have been found to differ in vitamin content. The majority of authors (Kezely et al. 1945, 1946; Devyatnin 1948; Djaparidze 1965; Maurinya and Berzinya-Berzite 1974) obtained data showing a higher content of ascorbic acid and a higher ascorbate oxidase activity in female individuals of many arborescent and herbaceous plants. These results suggest differences in the activity of oxidizing and reducing processes between female and male plants. However, with respect to other vitamins, no clear differences have been found between individuals of different sex (Djaparidze 1965).

Differences between sexes have been found by some investigators in regard to other biochemical indices. In particular, it was established that in hemp, female individuals can be distinguished by a higher content of water, carbohydrates

---

[1]A more recent study of the leaves of *Gingko biloba* trees showed that the peroxidase patterns are clearly either that of the male or of the female; there are no intermediate forms. (H-w Zhong, Z-h Yang, G-l Zhu and Z-x Cao (alias T.H. Tsao), Scientia Silva-sinicae 18: No. 1, 1982) [Ed.].

(Leisle and Makarova 1950; Herich and Priehradny 1955; Djaparidze 1965), calcium, nitrogen, potassium, and phosphorus (Dobrunov 1935; Erdelsky and Herich 1956; Yakushkina and Khrianin 1967). Erdelsky and Herich found differences in the distribution of potassium between various organs of the plant. They observed that in male plants the aerial part was richer in potassium than the roots, whereas the opposite was true in females.

It can be argued from the above data that, just as with morphological characters, the levels of activities of certain biochemical processes are characteristic of one sex or the other. To relate the form of sex determination to the activities of physiological processes, numerous comparative studies of the levels of photosynthesis, respiration, and transpiration of male and female plants have been undertaken. The results obtained lacked reproducibility because the ages of plants of different sex were not comparable, since male plants generally terminate their development faster than females.

The level of photosynthesis in plant ontogenesis is not constant throughout plant development; it varies as a function of time and is dependent on the phases and the rates of development of male and female individuals. In the first stages of individual development (i.e., before flowering), photosynthetic activity is generally higher in male forms. However, in later stages of vegetative growth (i.e., starting with flowering) the situation is reversed. This regularity is found in hemp (Dobrunov 1935; Valter et al. 1940; Khrianin 1964), spinach, sorrel, squill (Weiling 1940; Hartmann and Durand 1969), and in certain other dioecious plants (Chrelashvily 1941; Djaparidze 1965). However, comparison of the levels of respiration between male and female plants did not reveal any consistent differences (Minenkov 1924; Djaparidze 1941; Chrelashvily and Djaparidze 1950; Hartmann and Durand 1969). Detailed studies by Naugolnykh (1958) and Djaparidze (1965) on more than 20 species of dioecious plants showed that the differences in the levels of photosynthesis and respiration are not reliable markers for the characteristic activities of plants of different sex.

Inconsistent results were also obtained when the levels of transpiration were studied in groups of plants of different sex types. Moniava (1948) and Leisle and Makarova (1950) found higher rates of transpiration in female plants of persimmon, hemp, pistachio, and campion than in males. Moreover, the work of Djaparidze and Moniava (1948) showed that the high levels of transpiration were paralleled by higher water content in the tissues of female plants. In hemp, however, according to Kurilova (1935), the transpiration coefficient is higher in male individuals than in females. The experiments of other authors (Weiling 1940; Naugolnykh 1945) did not reveal any clear relationship between the level of transpiration and sex expression. However, the large body of experimental evidence on water content and the rate of water supply amply supports Djaparidze's statement (1965) that female dioecious plants tend to differ from males in their higher water content.

The contradictions obtained in numerous studies of physiological and biochemical processes can most probably be explained by the differences in the life cycles of male and female plants. The longer developmental cycle of female plants allows for a wider variation in the levels of physiological processes. As a result, during consecutive identical growth periods, female and male plants of the same chronological age are apt to be in different phases of their biological activity.

D. Foundations for a Biochemical Diagnosis of Sex

In the beginning of the 20th century, a great many papers appeared that dealt with the comparative study of biochemical peculiarities of male and female plants and animals. The differences in sexual characters thus were explained by the actions of sex hormones secreted by the endocrine glands (Steinach 1910, 1913; Pezard 1915, 1928; Zavadovskyi 1923), and were related to the levels of oxidation and reduction of cells and tissues. In this respect, interesting work was performed by Manoylov (1923a,b) who proposed, as a means to diagnose sex, the use of color reactions indicating the redox potential of tissues. In particular, he observed that blood taken from female humans or animals had a higher reducing capacity than blood from males. Initially, Manoylov (1923a) thought that the differences in rates of a given reaction depended on the presence in blood of specific sex hormones. Later, however, he concluded that the difference in the reaction must be caused by the predominance in the blood of males of easily oxidizable organic substances, and by differences in enzymatic activities and in hemoglobin content in the blood of male and female organisms. Popov (1926a,b), Schmidt and Perevozskaya (1926), and Galyalo et al. (1926) showed that the differences between the blood of males and females were related to differences in the content of proteins. Biochemical features characteristic of the two sexes were also revealed in a number of dioecious plants (Manoylov 1924). Extracts obtained from various parts of female plants showed a higher reducing power than similar extracts of male plants of the same species. Grunberg (1924a,b) applied Manoylov's reaction to many plant species (*Cannabis sativa*, *Hippophae rhamnoides*, *Urtica dioica*, *Begonia* sp., and *Populus* sp.), and in the majority of cases Manoylov's observation was confirmed. However, when Grunberg assayed Begonia flowers (i.e., tissues that lacked chlorophyll), he found a higher oxidative power in the stamens and a higher reducing power in the pistils. He therefore concluded that the reaction in plants did not depend on chlorophyll but probably on some hormonal substances.

Subsequently, detailed studies were conducted on higher plants by Satina and Blakeslee (1926, 1927) and on fungi by Ruzinov (1927a,b). A large number of species reacted as expected to Manoylov's assay. Only 5% to 7% of the plant species studied did not react as predicted. Correns (1928a,b) explained these and some other negative results found by Burgeff and Seybold (1927) by imprecision in the methods used to conduct Manoylov's assay; in particular, he emphasized the importance of the amounts of extract and reagent. Many researchers (Galyalo et al. 1926; Burgeff and Seybold 1927; Ryzhkov 1936) investigated Manoylov's reaction and came to the conclusion that the results of the reaction are not determined by qualitative differences, i.e., different compounds, but rather by quantitative differences in the rates of individual biochemical processes. Although Manoylov's assay did not become a universal means to diagnose sex, it did show that, in general, the tissues of male organisms are capable of highly intensive oxidative processes, whereas female tissues have a higher reducing potential.

Kizel (1940) wrote that Manoylov's observations and generalization have been confirmed in many instances in more recent work. In particular, the finding that male and female individuals differed in their redox potentials served as a basis for the physicochemical theory of sex formation developed by Joyet-Lavergne (1931). The first tenet of this theory is that cells that will develop into the gynaecium have a higher reducing capacity than cells that will form the androecium. Joyet-Lavergne believed that iodine-reducing substances (glutathione and others) were very important in controlling sex expression. Later studies conducted by Djaparidze (1965) on 9 plant species showed that the total content of iodine-reducing substances was 20% higher in the leaves of female individuals than in the leaves of males, which gave a more specific meaning to the generalization described above.

The second tenet of Joyet-Lavergne's theory is that female plant cells preferentially accumulate lipids and soluble carbohydrates; cells of the male plant, on the other hand, accumulate nitrogenous compounds. In other words, the type of cell metabolism indicates the direction of sex formation in a given organism.

Using histochemical techniques, Valter and Lilienstern (1934a,b) studied the sex-related differences in biochemical processes in the apical cone of hemp shoots. They concluded that the embryonic tissue of the apical cones had a higher oxidative power in male plants and a higher reducing power in female plants. Proceeding from Joyet-Lavergne's postulates, Sabinin (1940) proposed that if a given mineral nutrition was maintained, the determination of sexuality in plants was related to changes in the redox system of the cells and in the general biochemistry of the plant. In more recent times, detailed studies in this area were conducted by Maurinya and his co-workers (1974). They demonstrated that in corn and cucumber plants, sex determination and the type of redox processes were indeed interrelated. Female sex formation was linked to a lower redox potential in the tissues of generative organs, and the opposite was true for male sex formation.

Thus, considering all the preceding data, it can be concluded that there is a definite link between sex expression and the redox potential of plant tissues.

E. The Concept of Plant Sex Hormones

The discovery of animal sex hormones and the intensive studies that followed the publication of Brown-Séquard's experiments in 1889 showed very clearly the significance of sex hormones for vertebrate development. Detailed studies of the centers of hormone formation, of the distribution of hormones throughout the body, and of their effects, often provided a simple and clear explanation for many events in animal development (Trendelenburg 1932). The presence of sex hormones in animals suggested that such hormones possibly existed in plants as well.

As early as 1925, Berg proposed that there were plant hormones specific for the male and the female sex, and that these hormones were the same in higher plants as in higher animals. Later, Loewe and Spohr (1926) indicated that parts of flowers contained substances having estrogen-like activities. Butenandt and Jacobi (1933) confirmed that these substances were indeed present and showed that they were chemically identical to certain steroid hormones. Independently, one of us (Chailakhyan 1937) suggested that plant sex hormones participate in the transition from vegetative growth to flowering. The transition in male plants would involve the male sex hormone, in female plants the female sex hormone, and in monoecious plants both hormones.

In many animals, sex differentiation is under the control of sex hormones that are formed in the gonads. The male animal produces male sex hormones, or androgens. With testosterone as the major representative, androgens constitute a group of steroid hormones. Testosterone and related neutral 19-carbon steroids also represent a considerable fraction of the steroid hormones circulating in the female (Pokrovsky 1976). It has been established that testosterone can be transformed into estrogens.

In contrast to androgens, the female sex hormones or estrogens are not a homogeneous family of compounds, but are divided into two groups that differ in structure and biologic function: estrogens (major representative, 17-β-estradiol), and gestagens, or progestins (major representative, progesterone, Figure 1.1).

The estrogens that are present in the male organism apparently play an active role in the various processes of intracellular regulation (Pokrovsky 1976). Thus, male and female organisms are capable of synthesizing both androgens and estrogens, albeit the ratios of these hormones are different (The ratios can change with age too (Ed.)).

The role that animal sex hormones may play in plant physiology has been extensively studied by many research teams. The first review on this subject appeared in 1945 (Löve and Löve 1945). Early experiments on *Cannabis sativa* L. and *Mercurialis annua* (Orth 1934) did not reveal any influence of animal sex hormones on sex differentiation in plants. Later, however, experiments on *Melandrium rubrum* yielded some positive results (Löve and Löve 1940, 1945). The experiments consisted of coating decapitated stems (i.e., the apices, together with all the flower buds, had been removed) with a lanolin paste containing various sex hormones in a concentration range from 0.0005% to 0.1%. Estradiol, estrone, and estradiolbenzoate enhanced female sex organ formation, whereas testosterone and testosterone propionate enhanced male sex organ formation.[1] A positive influence of animal sex hormones on sex expression in plants was also demonstrated in the case of spinach (Hylmö 1941) and hemp (Kubare 1965). In

[1]i.e., pistillate and staminate flowers, respectively [Ed.].

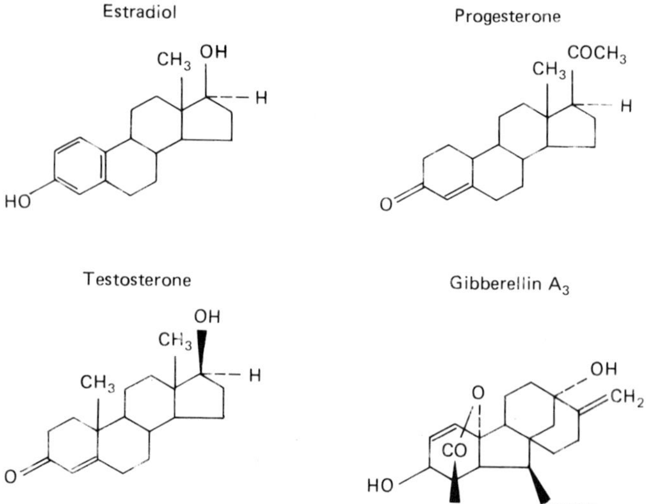

FIGURE 1.1. Comparative structures of the three steroids mentioned in the text and of gibberellic acid. The most important difference between the two types of steroids is the benzenoid ring in the estrogens (e.g., estradiol) as compared with a cyclohexenone ring (only one double bond) in progesterone. Comparison of the structural formulae shows that the configuration of gibberellin A_3 is more related to that of the androgens because neither has an aromatic ring [Ed.].

some other cases, however, the effect on plant sex formation was the same whether male or female animal sex hormones were used. For instance, the treatment of cucumber plants with either 17-β-estradiol or testosterone led to the formation of female sex organs. In summary, the studies on the influence of animal sex hormones on plant sex development showed that only in certain cases did male sex hormones cause male sex organ formation and female sex hormones cause female organ formation.

Steroid hormones have been discovered not only in vertebrate and some invertebrate animals but in plants as well. The female sex hormone estrone was found in palm fruits by Butenandt and Jacobi as long ago as 1933, and later was identified in date kernels and in the seeds of pomegranate and apple by Heftmann (1972). Estriol can be found in willow flower clusters, and androstanetriol in *Haplopappus heterophyllus* (Zalkow et al. 1964). More recently both estrone and estriol have been detected in plant tissues (Pokrovskyi 1976).

It was found in Perilla that estrogens, absent from plants growing under noninductive long-day conditions, appear in primordial inflorescences when these are induced to form by exposure to short days (Heftmann 1971). Similar results were obtained in experiments on goosefoot, *Chenopodium*. Only two short days (needed to induce flowering of goosefoot) led to an increase in estrogen content. Another example is provided by bean (*Phaseolus*) plants; initially the shoots do not contain any estrogen-like compounds, but these appear during the establishment of flowers, and their content reaches a maximum during flower-bud growth and seed formation. In cucumber, the increase in the level of endogenous estrogens coincides with the increase in the number of pistillate flowers (Kopcewicz and Chrominski 1972). An enzymatic system metabolizing estrogen has been discovered in bean plants; it was also noted that the leaves contained higher levels of estrogens than the stems (Joung et al. 1979).

In terms of chemical structure, the gibberellins and certain other phytohormones (Chailakhyan 1958, 1964, 1968; Chailakhyan et al. 1977) are distantly related to animal steroid hormones. Moreover, these hormones have common biosynthetic intermediates. Gibberellins are biosynthesized through the terpenoid pathway (Muromtsev and Agnistikova 1973),[1] and some cytokinins contain the terpenoid C5 unit (Paseshnichenko and Guseva 1974). Comparison of the structural formulae of gibberellins and animal sex hormones reveals that the configuration of gibberellins is somewhat closer to that of androgens (Fig. 1.1), for, in contrast to estrogen, neither gibberellins nor androgens contain an aromatic ring. It is interesting to note that cholesterol, which is the precursor of all animal steroid hormones, including sex hormones, is also present in many flowering plants (Paseshnichenko and Guseva 1974). Large amounts of cholesterol are found in the pollen of *Hynochoeris radicata*. The formation of steroid hormones from cholesterol in certain lower plants and in animals has the same intermediates, namely pregnenolone and progesterone (Heftmann 1972).

As yet, there is no reliable information to suggest that the modes of action of phytohormones and of animal sex hormones are identical (or even related (Ed.)). However, there are data indicating that there are interrelations between the two groups of substances. Thus, treating plants with estrogens leads to an increase in endogenous cytokinins but does not affect the level of abscisic acid (Kopcewicz and Rogozinska 1972). Correspondingly, kinetin increases the level of estrogens in bean plants, whereas gibberellins and abscisic acid have just the opposite effect (Kopcewicz 1972a,b). The actions of hormones in plants and the actions of steroid hormones in animals also have something in common, in that the action of phytohormones in plants and the action of hormones in animals are both related to changes in enzymatic activities and probably also in cytoplasmic membranes; both affect nucleic acid and protein synthesis; both have influences on the genome (Bonner 1967, 1968; Mueller et al. 1961; Khrianin 1969; Polevoy 1970; Heftmann 1971; Sergeev et al. 1971; Staroseltseva 1976; Kulaeva 1977; Muromtsev and Agnistikova 1973; Kefeli 1974; Gamburg 1976). Thus, the discovery of a number of steroid compounds in plants, their relationship to the flowering processes, and their similarity to animal steroid hormones (Figure 1.1) are all facts that favor the existence of sex hormones in plants.

---

[1]The biosynthesis has recently been summarized by B. O. Phinney: Biol. Plantarum 27, 172–179 (1985) [Ed.].

2
Influence of Environmental Factors and Nutrition on Sex Determination in Plants (A Review)

The practical and theoretical aspects of the problem of sex regulation by external factors have been of great interest to many botanists. The answers to these questions will increase our ability to understand and direct the growth and development of plants. The determination of sex and the formation of sex organs during growth and development are determined not only by the genetic apparatus, but also by environmental factors. There is a considerable body of evidence indicating that external forces cause changes of sex characteristics in animals and plants, under both natural and experimental conditions (Pezard 1915; Zavadovskyi 1923; Morgan 1928; Grishko 1935; Astaurov 1937, 1940, 1963; Minina 1952; Heslop-Harrison 1957; Strunnikov 1962; Djaparidze 1963, 1965; Lvova 1963; Vince-Prue 1975; Frankel and Galun 1977). As yet, all the extensive experimental material and theoretical considerations have failed to give a definite answer to the question of exactly how environmental conditions act to change the sex in different plant groups. The extent to which these sex transformations involve solely genetic control (Correns 1928), or solely metabolic changes (Schaffner 1927), or both (Sabinin 1940) remains an open question.

One of the earliest researchers to have described the control of sex formation by environmental conditions was the Russian botanist I. A. Dvigubskyi (1823). Dvigubskyi showed that carob plants grown in Europe had unisexual flowers, whereas those grown in Egypt had hermaphroditic flowers.

A. Mineral Nutrition and Water Supply

The influence of mineral nutrition on sex transformation and the related changes in the ratios of male and female flowers in different plants have been extensively surveyed (Molliard 1897; Schaffner 1921; Riede 1922; Gardner 1922; Correns 1928; Minina 1935). The first study addressed the general effect of poor and rich soils on sex expression of plants (Molliard 1897). Using hemp, mercury, begonia, and other plant species, the majority of authors (Heyer 1884; Hoffman 1885; Geddes and Thompson 1889; Atkinson 1898; Giard 1898; Halsted 1901; Sprecher 1913; Davey and Gibson 1917; Schaffner 1922, 1925a,b,c; Maekawa

1924, 1927, 1929; Camp 1932; Mukerye 1936; Matzke 1938) obtained results indicating unambiguously that rich soils enhanced female organ formation and poor soils enhanced male organ formation.

Subsequent studies of sex change in relation to nutrition addressed the more specific question of the role of individual mineral nutrients. As early as 1933, Sabinin (1934) suggested that fertilizers could affect sex change in plants. Specifically, he proposed that individual elements and their relative concentrations in the soil could play an important role in determining the direction of sex expression (Sabinin 1940, 1971). This hypothesis led to the series of studies conducted from 1935 to 1940 by Sabinin's co-workers, E. G. Minina, V. A. Gusieva, F. Z. Borodulina, and N. A. Satarova. The effects of fertilizers on plant sexual characteristics were studied by growing plants under varied conditions of mineral nutrition; nitrogenous and potassium fertilizers were added periodically (at the times of seeding, budding, and flowering). Periodic additions of nitrogenous fertilizers resulted in a considerably greater number of female inflorescences in corn; for every male inflorescence (tassel) 5 to 6 female inflorescences (cobs) were formed. Potassium fertilizers had the opposite effect: the inflorescences were predominantly of the male type (Minina 1935, 1936; Minina and Guseva 1937). The experiments of Kh. A. Maurinya (1956, 1961, 1963) gave analogous results. Periodic addition of nitrogenous compounds to the growth medium enhanced the formation of female flowers in corn, and was accompanied by a characteristic decrease in the redox potential and lower viability of the pollen. Addition of potassium compounds, on the other hand, resulted in an increase in the number of male tassels, a high viability of pollen grains, and a high redox potential (E_h = 120 mV) characteristic of tissues of male plants.

According to Sabinin (1940) the results obtained by Minina et al. are not to be explained by a specific response of corn to the nitrogen nutrition, but rather they were a consequence of the general effect of nutritive conditions on sex differentiation. The conditions of nitrogen nutrition drastically affected the ratios of individuals of different sexes in pumpkin plants as well (Minina and Guseva 1937; Borodulina 1938). Experiments in the greenhouse and in the field showed that moderate nitrogen starvation during the establishment of reproductive organs leads to a change in the ratios of female to male flowers; the number of female flowers is increased. It should be noted that in cucumber, the numerical ratio of male to female flowers is normally stable and characteristic for a given cultivar. The closer to the base of the stem a female cucumber flower is borne, the higher the expression of the female sex in that flower (Pyzhenkov 1968).[1] In Satarova's experiments on melon plants (1936), periodic additions of nitrogen increased the number of female flowers more than twofold compared with the controls. This naturally led to an increased harvest. Periodic additions of potas-

[1]This contradicts not only the data with melon immediately following, but also the extensive results of Nitsch et al. (Am. J. Botan. 39:32–39, 1952), which showed that male flowers are formed first and then, in succession, normal female flowers and "superfemales," i.e., parthenogenetic females [Ed.].

sium led to a decrease in the number of pistillate flowers and delayed their opening by 3 to 5 days compared with the controls. In every case, periodic nitrogen addition increased the overall rate of development of the plants as well as the number of female flowers. The results obtained in these experiments have led to a general recommendation to use high nitrogen nutrition in selecting the fertilizer; this is now considered a general rule for the cultivation of vegetables.

The experiments of Minina and Guseva (1937) showed that periodic additions of nitrogen can more than double the amount of stored sugars and decrease the total nitrogen content of the tissues of cucumber shoots of the first and second order. Based on these data, Sabinin (1940) came to the conclusion that the method of nitrogen nutrition used, while insuring an adequate level of photosynthesis, limits synthesis of protein. Because sugars are reducing substances, it follows that, with a given nitrogen supply, the ratio of reducing to oxidizing processes has been displaced in favor of the reducing processes. According to the theory of Joyet-Lavergne, such a displacement results in an enhancement of female sex organ formation, which was indeed observed in the experiments of Minina et al. Subsequently, Naugolnykh (1948) showed that presoaking cucumber seeds in a solution of methylene blue led to an increase in the number of pistillate flowers; this was accompanied by storage of large amounts of sugars and a decreased oxidizing potential of the plant cells. According to Molotkovskyi (1965, 1974), the area where female inflorescences of corn are formed (from the root collar to the site of attachment of the uppermost cob) is rich in proteins, while the area where male inflorescences are formed (from the internode above the uppermost cob to the top of the panicle) is rich in carbohydrates.

Thus, Sabinin's co-workers have shown that individual elements in mineral fertilizers and the time when they were supplied play specific roles in sex determination. Other researchers obtained largely parallel results that have emphasized the importance of nitrogen nutrition. In early experiments on hemp grown on Knop's solution, high levels of nitrogen determined female sex expression, and low nitrogen levels determined male sex expression (Tibeau 1936). Growing spinach on Hoagland's solution containing various levels of nitrogen, Thompson (1955) observed that high nitrogen content in the solution resulted in an increase in the number of female and monoecious plants.

Both in solution culture and in soils, the influence of nitrogenous compounds on sex expression was modified when the plants were grown under different day-lengths. Tiedjens (1928) observed that when cucumber plants (*Cucumis sativus*) were grown in nitrate-rich soil under long-days, the number of staminate flowers increased by 45.5%, and the number of pistillate flowers by 55.08%. Under short-days, however, the increases in the numbers of flowers were, respectively, 3.1% and 20.7%. In a similar set of experiments, Hall (1949) grew cucumber plants (*Cucumis anguria*) under different day-lengths in nutrient solutions containing varied nitrogen levels. With high levels of ammonium nitrogen, the ratio of male to female plants was 4.58:1 under short-days (8 hours of daylight) and 4.87:1 under long-days (16 hours of artificial light). When only low levels of nitrogen were present, and the plants were given long days, this ratio was 6.0:1.

Thus, the enhancement of female sex expression by high nitrogen levels is somewhat stimulated by short-day growth conditions.[1]

Increased nitrogen content in soil combined with short-day length resulted in the formation of pistillate flowers in the male tassels on the tips of corn plants (Choudhri and Krishan 1946). Long-day lengths had just the opposite effect. Experiments on plants that bear hermaphroditic flowers, such as tomatoes, showed that high nitrogen levels in soil led, under short day-conditions, to pertubations in meiosis, to the formation of sterile pollen, to the reduction of stamens, and to the transformation of stamens into pistils (Howlett 1936, 1939).

Thus, by changing certain conditions of mineral nutrition, it is possible to regulate sex expression in dioecious and monoecious plants. The influence of some elements in mineral nutrition on sex formation of either type can be explained by peculiarities in the growth of male and female plants of dioecious species, and by the fact that they may require different amounts of a given nutrient. Also, in the majority of dioecious species the duration of vegetative growth of the female plants is considerably longer than that of the males. This suggests that the absorption of nutrients from the soil may be different for individuals of different sexes. Indeed, Dobrunov (1935a,b) showed that, from initial developmental stages up to flowering, male hemp plants absorb nutrients more efficiently than females, while after flowering the rate of absorption of nutrients and the accumulation of dry matter is higher in female plants.[2]

An important factor in sex expression in plants is water supply. The work of Djaparidze (1945, 1965) showed that female individuals always differ from males in their higher water content. Moreover, female reproductive organs and gametophytes are more resistant to excess water in soil than are males (Shazkin et al. 1968). Hence, it was expected that increased water supply would favor the establishment and differentiation of female flowers, whereas insufficient water supply would do the same for male flowers. Indeed, it was demonstrated that high moisture in soil and air during the differentiation of the initials of generative organs is a necessary condition for the establishment and development of female characteristics; dryer conditions are required for male sexual characteristics (Minina 1952). These experiments of Minina (1952) also showed that if cucumbers are grown under high moisture conditions, the ratio of staminate to pistillate flowers is lowered by a factor of 10, compared with that of plants growing under low moisture conditions. In cucumber plants grown in polyethylene greenhouses (relative humidity in air: 96% to 100%) with soil moisture at 80% of maximum water content, Pashenko and Redman (1968) noticed the appearance of a considerable number of hermaphroditic flowers, a massive proliferation of ovaries, and even the formation of female flowers on tendrils. Soil moisture at 60% to 80% of its maximum favored the appearance of monoecism in hemp (Arinshtein

[1]However, the conclusion is weakened by the fact that there was a predominance of male plants in spite of the high nitrogen [Ed.].

[2]This coincides with the growth and development of fruits and seeds [Ed.].

and Loseva 1958). All these experiments show that high water contents, both in the substrate and in the atmosphere, are factors contributing to the establishment and differentiation of female flowers and, in dioecious species, to the formation of female plants.

B. Atmospheric Gas Composition

An important role in plant sex determination is played by the composition of the atmospheric gas. As early as the 1860s, market gardeners from Klin developed the so-called method of "curing" cucumber seedlings. This method consisted of fumigating cucumber plants of a certain age with gases resulting from incomplete combustion of firewood. Plants were usually treated for two 12-hour periods. The effects of this treatment could be seen for up to a month later. The fumigation resulted in a series of morphological and physiological changes, including epinasty, contortion of leaf blades, a decrease in chlorophyll content, and decreased water retention capacity of the tissues. This "curing" of plants with carbon monoxide (no doubt mixed with ethylene (Ed.)) led to an early appearance of pistillate flowers, increased their number, and generally improved the subsequent harvest. The experiments of Minina and Tylkina (1947) demonstrated that the change of sex that followed the fumigation of cucumber plants by this method was related to the fact that incomplete combustion of wood led to the formation of carbon monoxide and ethylene. The authors conducted detailed studies of the influence of these gases on sex formation in cucumber, corn, spinach, and wild strawberry. Experimental plants (at the 2–3 leaf stage) were cultivated for 50 to 200 hours under a bell jar with either gas; the plants were aerated daily for 2 to 3 hours and the gas supply renewed. Such treatment led to an increased formation of female flowers. In particular, cucumber plants exposed to 1% CO carried female flowers only. The increase in female sex formation also occurred under the action of CO, 0.5% to 0.3% plus ethylene, 0.3%. Ten days after the treatment of spinach with 1% CO, male plants developed a large number of leaves, which is characteristic of female plants, and flowering was delayed by 5 days.

Gases like these can alter the functioning of the redox system of the cell. Oxidative processes are inhibited, and this is accompanied by almost halving the content of ascorbic acid, especially in its reduced form (Minina 1952). Changes in the redox system that take place under the action of these gases thus contribute to the formation of female sexual characteristics (Minina 1949; Minina and Kushnirenko 1949). It was established in subsequent investigations that the treatment of hemp with 1% CO caused an increase in the number of pistillate flowers in comparison to staminate flowers. Experimental male plants developed mixed flowers. The stamens of male flowers carried stigmalike formations; in other cases the thickening and the fusion of stamens led to structures reminiscent of a normal pistil with an ovary. Subsequently, CO and ethylene were also found to influence sex expression in *Scilla* (Heslop-Harrison and Heslop-Harrison 1957;

Heslop-Harrison 1957, 1959). A direct and very marked influence of ethylene on female sex formation in corn was observed by Molotovskyi (1940). Among ethylene-producing compounds, Ethrel (2-chloroethylphosphonic acid) has been the most widely used. This compound is converted to ethylene in plant tissues. Treatment of cucumber plants with Ethrel at a concentration of 240 ppm led to the formation of exclusively female flowers on those nodes that normally carry male flowers (McMurray and Miller 1968; Robinson et al. 1969; Kumarasamy 1972).

In recent years a number of publications have appeared in which it was shown that Ethrel stimulated the development of pistillate flowers and inhibited the formation of staminate flowers in many species of *Cucurbitaceae* (Sims et al. 1970; Iwahori et al. 1969, 1970; Rudich et al. 1969, 1970; Karchi 1970; Freytag et al. 1970; Sustikova and Ginterova 1973; Minina and Larionova 1979; Tarakanov and Agapova 1973; Agapova 1975; Moursy and Khalil 1976). In experiments of Tronichkova (1978), a single Ethrel treatment of cucumber plants reduced the number of male flowers; a repeat treatment completely prevented the development of any male flowers. Such a method could be used in the production of hybrid cucumber seeds. A similar method has also been developed for melon (Sankin et al. 1978).

The effects of ethylene on sex expression have been demonstrated in other plants as well. The spraying of hemp with Ethrel led to the formation of female flowers on male plants (Mohan, Ram, and Jaiswal 1970; Davidyan and Rumyantseva 1974). Treatment of grape vines at the stage of bud break with Ethrel resulted in the development of fertile hermaphroditic flowers on the main shoot which normally carries only staminate flowers (Kender and Ramaily 1970). Ethrel also accelerates the development of the pistil in *Cleome spinosa* (Jong and Bruinsma 1974b). The use of Ethrel on graminaceous plants (wheat and barley) inhibited the differentiation of the anthers and led to male sterility (Bennet and Hughes 1972; Hughes et al. 1974; Law and Stoskopf 1973).

It has also been shown that the treatment of *Cucurbitaceae* with acetylene, particularly at the stage of 1 to 2 true leaves, results in an increase of up to 23% in the number of pistillate flowers (Mekhanik 1958).

Some authors believe that the Ethrel-specific inhibition of staminate flower formation in the gourd family is due to the effect of Ethrel on endogenous ethylene content, and does not involve the conversion of Ethrel into ethylene (e.g., Loy 1971). Similarly, it has been suggested that endogenous ethylene plays a regulatory role in sex differentiation in *Cucurbitaceae* (Byers et al. 1972). Endogenous ethylene is synthesized from methionine (Lieberman 1975).[1]

[1]By way of S-adenosyl-methionine and 1-aminocyclopropane-1-carboxylic acid or ACC (Adams and Yang: Proc. Natl. Acad. Sci. USA 76:170–174, 1979; Lürssen et al.: Zeit Pflanzenphysiol. 92:285–294, 1979). The remarkable ability of ethylene to stimulate the flowering and therefore the fruiting of the pineapple and lichee nut can be thought of as an extension of this trend toward femaleness [Ed.].

The extensive body of experimental evidence thus clearly demonstrates that certain gases, namely carbon monoxide and ethylene, do influence sex expression in plants.

C. Temperature

Many researchers, in the course of their studies of the effects of various factors on sex expression in plants have simultaneously determined the influence of temperature on this process.

Nitsch et al. (1952), in particular, conducted experiments on pumpkin (*Cucurbita pepo*) in the rigorously controlled environment of a phytotron. They demonstrated that long days and high temperatures enhance the formation of male flowers, whereas short days and low temperatures favor formation of female flowers. Similar results were obtained with giant ragweed (*Ambrosia trifida*) by Jones (1947), and later with spinach (*Spinacia oleraceae*) by Thompson (1955). V. V. Anisimov (1966, 1967) noticed that in hemp, relatively low temperatures (10–12°C) induced proliferation of flower clusters, and that the sex of the flowers changed. Anisimov (1967) believes that during the brief time in which the plants are sensitive to photoperiod, low temperatures strongly inhibit the processes of differentiation triggered by short-day conditions, and also that they alter the metabolism.[1]

Experiments on hemp (Nelson 1944; Heslop-Harrison 1972) showed that a decrease in temperatures not only results in a higher proportion of female plants in the population, but that it also stimulates the growth of female flowers on male plants. In view of these results, Frankel and Galun (1977) suggested that changes in sex regulation might occur at different stages of development. Thus a decrease in temperature at an early stage would lead to the transformation of a genetically male individual into a phenotypically female plant, while at later stages the sex change would be confined to individual flower buds. This would result in the formation of monoecious plants. Finally, low temperatures at very late developmental stages would lead to the formation of bisexual flowers. Indeed, such bisexual flowers have been obtained in many dioecious plants grown under defined temperature conditions (Nelson 1944; Jones 1947; among others). Grishko (1935) showed that a decrease in temperature during budding leads to the formation of bisexual flowers on male hemp plants. The flowers were of the staminate type, but carried a small number of pistils. Female sex expression was also induced by low temperatures in male papaya plants (*Carica papaya*) (Lang 1961). On the other hand, insufficient exposure to cold during flower induction in olive trees results in the appearance of predominantly male flowers with rudimentary pistils (Badr and Hartmann 1971). Apparently, temperature fluctuations during flower

[1]This tendency of short days to favor femaleness was noted in the experiments of Tjedjens (1928) and Hall (1949) [Ed.].

induction affect primarily the formation of flower parts and sex expression. The work of Medvedeva (1933) demonstrated that low temperatures and frosts during flowering of hemp results in a series of perturbations in meiosis, which in turn lead to the formation of pollen grains (male gametes) with an altered chromosomal set.

However, low temperatures do not favor female sex formation in all dioecious plants. For example, in one species of mercury (*Mercurialis annua*) high temperatures led to female sex expression, and low temperatures led to male sex expression (Molliard 1898b).[1] The experiments of Stau on hyacinth (cited by Lvova 1963) are of particular interest. Stau observed that under the influence of high temperatures, pollen precursor cells give rise to structures analogous to embryonic sacs. When these structures and normal pollen were grown together on artificial media the pollen tubes grew directly towards these embryo-like formations. Thus high temperatures not only altered the morphological characters of the pollen grains, but shifted their biochemical properties toward those of the female type. Similar changes have been observed in the anthers of star-of-Bethlehem (*Ornithogalum umbellatum*) and potato.

It should be noted that the formation of female individuals under the influence of high temperatures can be observed in the animal world as well. For instance, Astaurov (1937, 1940) found that temperature activation of virgin reproduction in mulberry silkworms gives rise to an almost exclusively female progeny ("temperature-dependent parthenogenesis").

Frankel and Galun (1977) studied the influence of temperature on monoecious plants carrying unisexual flowers (corn and cucumber) and discovered that at low temperatures, female flowers are developed at sites that in plants grown at higher temperatures would be occupied by male flowers. Howlett (1939) cultivated tomatoes during the winter months and noticed that the ratio of pistils to stamens was higher than in the control plants grown at other seasons. He believed this result was due to the short length of winter days, and that temperature played a rather insignificant role. However, other researchers (Smith, 1932; Osborne and Went 1953) have found that low temperatures alone stimulate pistil development and inhibit the functional formation of pollen, occasionally causing parthenocarpy. Rylski (1973) discovered the presence of overdeveloped and deformed ovaries in flowers that are formed at low night temperatures (8–10°C). Analogous results were obtained with experiments on eggplant (Northmann and Köller 1975). In banana plants, low temperatures (12°C) induce floral differentiation; embryonic bisexual flowers develop into imperfect female flowers that lack one or more carpels (Fahn et al. 1961).

These results thus indicate that high and low temperatures may have a drastic effect on the reproductive organs in the flower buds of a number of dioecious and monoecious plants, low temperature generally favoring femaleness.

---

[1]But male flowers of *Mercurialis* often have an ovary, though it is nonfunctional, i.e., they are only partially male [Ed.].

There is some evidence regarding the influence of temperature on sex formation in grasses, but here the changes in sex expression in reproductive parts occur at much later developmental stages (Frankel and Galun 1977). It has been shown in barley that an extremely hot period in the beginning of the summer is followed by a teratological development of both anther and pistil (Gregory and Purvis 1947). Wheat grown at low temperatures in constant light gives rise to some anthers having the structure of a carpel (Meletti 1961). However, these and other results (Meyer 1966; Napp-Zinn 1967) have also been obtained under natural environmental conditions where temperature was not the most important variable.

Pleshakov (1951) discovered that warming dry cucumber seeds results in an increase in the number of pistillate flowers. Female sex expression is favored by changing the temperature conditions: cucumber seeds can be chilled to 0–2°C and then frozen at −2°C and −5°C (Vladimirova 1952); alternatively, the seeds can be warmed to 37°C and then chilled to 2°C (Lvova 1963). Such treatments of cucumber seeds before sowing change the pathways of physiological and biochemical processes: the contents of reducing sugars and ascorbic acid are increased. This results in an overall increase in the reducing power of the cells of the seedling, which in turn favors female sex expression (Kandina 1958).

According to Lvova (1963) increases in temperature lead to an earlier differentiation of the apical cone than in control plants. The early developmental stages proceed at a much more rapid rate, resulting in an early appearance of female flower initials. This last fact explains the formation of a larger number of pistillate flowers than of staminate flowers. Thus, the shift in the ratio of female to male flowers of cucumber that is induced by certain temperatures can be explained by changes at a very early developmental stage, by alterations in the plant biochemistry and in the rates of differentiation of flower initials in the apical cone.

Providing that normal growth and development are made possible by the environmental conditions, low temperatures favor female sex expression in many dioecious and monoecious plant species. High temperatures have the opposite effect. The mode of action of the temperature factor on plant sex expression is probably related in some way to changes in metabolism (Chekmin et al. 1964; Dadykin 1952).

D. Day Length and Quality of Light

Day length, or photoperiod, is a factor of paramount importance in determining the shift from vegetative growth to sexual reproduction (Garner and Allard 1920). After the discovery of photoperiodism many researchers became interested in the influence that day length might have on sex expression in plants.

Photoperiodism plays a determining role in the control of sex formation in many plants (Minina and Larionova 1979) and in the change of sexual characteristics of animals (Belyaev et al. 1963). As early as in 1898, Molliard (1898a,b) and Strasburger (1900) noticed that when hemp and mercury were grown in a greenhouse in winter (short-day), the male plants carried some bisexual flowers. Tournois (1911, 1912, 1914) was the first to design specific experiments that demonstrated the occurrence of photoperiodism. He controlled day length by covering and uncovering the plants with black boxes. Tournois' conclusion was that short-day shifted the usual ratio (1:1) of male to female plants in favor of females (1:1.6) in the case of hop (*Humulus japonicus*) and hemp (*Cannabis sativa*). Moreover, he noticed that all 30 male hop plants carried a certain number of stamens having the structure of a carpel. In early experiments on *Silene noctiflora* conducted by Feuchting (cited by Lvova 1963), growth in insufficient light led to changes in the structure of the flower sexual elements.

The work of J. and I. Heslop-Harrison (1958a,b) is of particular interest. They determined that the reaction to the photoperiodic impulse could be mediated not only by the leaf (Moshkov 1936; Chailakhyan 1936; Psarev 1936), but also by the stamens and the pistil separately. An artificial change of the natural photoperiod in long-day greenhouse Silene plants (*Silene pendula* L.) results in deep morphological changes in flower structure. Moreover, the course of macro- and microsporogenesis is altered, and the flowers formed are unisexual, not bisexual. Plants react to changes in photoperiodic conditions in a variety of ways. When unfavorable photoperiodic conditions are applied, different photoperiodic plant groups undergo various phenotypic changes, including the time of flowering and the arrangement and morphological characteristics of flowers. In this respect, extensive studies have been conducted on hemp (*Cannabis sativa*) (Schaffner 1918–1935; McPhee 1924; Valter and Lilienshtern 1934; Breslavets 1933, 1936; Grishko 1935; Levchenko 1937; Borthwick and Sally 1954; Rudenko 1956; Arinshtein and Loseva 1958; Makarevich 1935, 1953, 1959; Heslop-Harrison 1957; Sironval 1959; Davidyan 1963, 1967; Anisimov 1967; Migal and Zhatov 1969; Khrianin and Chailakhyan 1977). The results of the majority of the experiments show that either short-day conditions or a decrease in the intensity of light induce a rapid reproductive development, inhibit growth, and lead to the appearance on male plants of bisexual flowers or flowers having predominantly female characteristics. Some of these flowers even set seeds, providing that short-day conditions were maintained over a long enough period. Long-day conditions have just the opposite effect.

In contrast to hemp, the long-day species of spinach react to short-day conditions by forming male flowers on genetically female plants. Thus, Thompson (1955) grew spinach plants in normal long days (15–16 hours) and obtained the following distribution of sexes: 48.4% females, 49.4% males, and 2.2% monoecious plants. In short days the distribution was changed to 28.3% females, 44.6% males, and 27.1% monoecious plants. However, earlier experiments (Rosa 1925; Knott 1932; Magruder and Allard 1937) yielded contradictory results. In some cases short days enhanced maleness, in others femaleness. As of today, the influence of short days on long-day plant species remains a controversial issue. In monoecious plants with unisexual flowers (corn, in particular), long days lead to the formation of male tassels in the axils of lower leaves, while short days stimulate the formation of ears on the apex of the plants (Schaffner 1927a,b). A short (7-hour) photoperiod, on the other hand, enhances the formation of bisexual flowers in staminate flower clusters of corn (Choudhri and Krishan 1946). Long days (18–24 hours) tend to inhibit female sex expression. A maximal increase of the photoperiod induces vegetative proliferation of corn male flower clusters, whereas short days ensure normal development (Galinat and Naylor 1951).

Experiments conducted on crops of the gourd family also showed that short days stimulate female sex expression (Tjedjens 1928; Edmond 1930; Whitaker 1931; Currence 1932; Hall 1949; Nitsch et al. 1952; Lvova 1963; Frankel and Galun 1977). Specifically, short day conditions in melon and cucumber plants induced the early appearance of pistillate flowers and an increase in their number. It is interesting to note that exposure of cucumber plants of a female race to extremely long days resulted in a larger change in the number of flowers (ratio of male to female flowers, 1.8:1) than in a male race (ratio of male to female flowers, 25:1). Heide (1969) showed that in Begonia, increased daily illumination enhanced female sex formation.

A number of experiments have been conducted on cocklebur (*Xanthium* spp). Growth of a short-day species (*Xanthium pennsylvanicum*) under long-day conditions led to the formation of male flowers in the experiments of Neidle (1938) and Naylor (1941). Interestingly, this result could only be obtained if the plants were first allowed to reach the flowering stage. However, Witsch (1961), in his studies on day-neutral species of cocklebur, did not observe any relation between retardation or acceleration of flowering and the final sex expression. This contradiction can perhaps be explained by the fact that not only did the different researchers use different species of cocklebur but they altered the photoperiod at different developmental stages; thus, the sensitivity of flower buds to environmental conditions may have been different in the different laboratories (Heslop-Harrison 1972; Khrianin and Milyaeva 1977). In response to environmental conditions, the rates of development of the flower parts can change (Correns 1928a,b), and this, in turn, can lead to change of sex (Gardner et al. 1934).

The type of sexual differentiation is influenced not only by the length of the photoperiod but by the spectral composition of the light as well. According to Sabinin (1955) and to Minina and Larionova (1979), the increased formation of male flowers in shoots of the lower part of the crown of trees is at least partially due to the fact that blue light penetrates deeper into the crown than red light. However, no common regularity could be found among many arboreal and graminaceous plants. For instance, the data of Andreenko and Kuperman (1959) on corn show that when the light is enriched in the long wavelength (red) part of the spectrum, the formation of female generative organs is retarded. Enrichment in the short wavelength (blue) part of the spectrum has the opposite effect: formation of female generative organs is stimulated. According to these authors, the formation of the panicle of corn before the ear is due to the fact that under long-day conditions the spectrum is rich in red light and also that as a result the development of the gynaecium lags behind the development of the androecium. Exposure to 9-hour days or to blue light more often affects the generative organs of functionally unisexual male flower clusters (tassels) of corn than those of female flower clusters or ears (Ustinova 1956). Short wavelength light stimulates female sex expression. It was shown with different varieties of cucumber (*cv.* Vyaznyakovskyi-37, Nyerosymye) that in 12-hour days (excluding morning and evening hours because they are rich in long wavelength light) the differentiation of carpel primordia was accelerated, while that of stamen primordia was slowed down (Lvova and Bakhanova 1961, 1962; Lvova 1963a). Such light conditions hasten female sex expression, and pistillate flowers appear in the axils of the third and fourth leaves, instead of after the sixth and seventh leaves, as is the case when plants are grown under natural 16-hour days (Vashenko 1959). In the same direction, Eisuke et al. (1968) determined that for the cucumber variety Higan-fishinari 8-hour days and blue light stimulated the formation of male flowers on lower internodes of the stem, while female sex expression was inhibited. Red light resulted in the formation of predominantly vegetative buds. Engelsman (1968) found that, under short-day conditions, a 1-minute burst of red light during the dark period results in an increase in the number of female flowers. Extending the treatment with red light leads to enhanced female sex expression. In photoperiodism a light break in the dark period is normally equivalent to giving a long day. However, when red light was replaced by far red, this stimulation was reversed.

These and later findings have led to the belief that sex regulation in unisexual plants is controlled (*inter alia*) by the phytochrome system. Changes in the spectral composition of light result in conformational changes in the phytochrome (Eisuke and Eiji 1970). Thus, in the red wavelengths the phytochrome is in its active form (P_{730}), while in far red light it is in its inactive form (P_{660}). Phytomorphogenetic processes presumably depend on plant metabolism, which in turn is influenced by red and blue light (Voskresenskaya 1965, 1975, 1979; Konstantinova et al. 1975).

These few data suggest that the influence of the spectral composition of light on the type of sex formation occurs through the phytochrome system and is probably related to the content and the activity of phytohormones.

E. Photoperiodic Induction and Darkness

To elucidate further the influence of light on sex expression, the authors of this book have studied the effects on sex expression of short-day and long-day photoperiodic induction, as well as of uninterrupted darkness.

The objects of the investigations were chosen with special reference to their biological peculiarities. The experiments presented here and in the following chapters were conducted on dioecious plants that have characteristic photoperiodic reactions. One of the main objects chosen was a short-day hemp species of the US-6 variety (*cv.* Yuzhnaya Sozrevayushaya). This variety gives a quantitative photoperiodic reaction and flowers both under short-day (SD) and long-day (LD) conditions; however, the flowering occurs much later in LD. A few experiments were carried out on other varieties of hemp: namely, *cv.* Starooskolskaya Ulutshenaya (SOU) and *cv.* Yuzhnaya Arkhonskaya.

Hemp (*Cannabis sativa* L.) belongs to the family *Cannabinaceae*. It is a typical dioecious plant. The characteristic feature of hemp is its high lability. It is therefore no coincidence that, among dioecious plants, hemp has been the most studied in terms of the impact of various environmental conditions on changes in sexual differentiation.

The other object of our studies was spinach (*cv.* Victoria). Spinach (*Spinacia oleracea* L.) is a dioecious plant belonging to the family *Chenopodiaceae*. It is an obligate LD species, i.e., it flowers in LD but not in SD. As in hemp, sexual dimorphism in spinach is clearly expressed. Male and female plants differ in their rates of growth and development and in many other physiological and morphological characters. Male individuals are less leafy than females and have a shortened life span.

Experiments were conducted both in the field and in the greenhouse. The field studies were conducted with usual culture conditions in the absence of fertilizers, in natural day length (location: Botanical Garden of the Penzenskyi Pedagogical Institute). Each experiment was repeated four times. Short 8-hour days were simulated in the field by covering the plants from 4 PM to 8 AM with plywood cells painted white. Greenhouse experiments were conducted in the greenhouses and growth chambers of the Timiryazev Plant Physiology Institute of the Academy of Sciences of the USSR. Plants were either grown in vessels containing Knop's solution or in boxes or pots containing soil. LD (16–18 hours) was composed of natural daylight supplemented (in winter, spring, and autumn) with artificial light (xenon arc water-cooled lamps, type: DKSTV 6000); SD (8 hours) and continuous darkness were created with the help of opaque screens. The temperature in the greenhouse was kept at 24–26°C. In the growth chamber, the plants were illuminated with type LBC-80 lamps, the temperature was kept at 20–22°C, and the humidity at 80%. The main goal of the experiments was to determine the optimal number of SDs and LDs necessary for change of sexual characteristics of dioecious plants. In the field experiments, young plantlets were separated into sets that were grown under the following conditions: 1) for hemp: a) constant SD, b) constant LD, c) 10 LD + SD, d) 20 LD + SD, e) 30 LD + SD; 2) for spinach: a) constant LD, b) constant SD, c) 10 SD + LD, d) 20 SD + LD, e) 30 SD + LD.

One of the results obtained was that in 8-hour days, although the growth of hemp was retarded, the plants started flowering 16 days earlier than under LD conditions. The number of female plants was increased, and some plants appeared that were intersexes, i.e., plants bearing flowers that have both male and female elements (cf. Goldschmidt 1915, 1920). Equalization of the numbers of male and female plants was observed where plants initially received 10 and 20 long days (10 LD + SD and 20 LD + SD), and the sex ratios were also equal in the 30 LD + SD group and the LD group.

The growth of spinach plants was also retarded under SD conditions, and the plants failed to flower. Flowering of the plants that initially received 30 short days

(30 SD + LD) was retarded by 23 days compared with the LD control. An increase in the number of female plants was only observed in this 30 SD + LD group (in the LD control the numbers of females and males were equal). A similar experiment conducted in a growth chamber showed that the increase in the number of female individuals was more dramatic if spinach plants were initially exposed to 40 short days.

Thus, sex formation in a short-day species of hemp and in a long-day species of spinach is under the influence of day length: an increase in the number of short days brings about enhanced female sex expression. It is well known that the night period plays a significant role in plant development: dark reactions are as important as light reactions. When some SD species—red perilla, cocklebur, beggarticks (*Bidens*), and hemp—are kept in constant darkness before being exposed to optimal day length, growth is retarded and flowering occurs (except in the case of hemp noted earlier (Chailakhyan 1943; Chailakhyan et al. 1970). Flowering is also stimulated by constant darkness in the case of the LD plant coneflower (Chailakhyan et al. 1976, 1977).

It was believed (Chailakhyan 1943) that darkness creates in the leaves of plants an internal physiological state that would favor the rapid formation of metabolites necessary for flowering, when favorable day length conditions are subsequently introduced. Thus, it became of interest to determine the effect of shorter exposures to continuous darkness on sex expression in hemp. Hemp plants (US-6 variety) were grown in boxes with soil on 18-hour days (at the Institute of Plant Physiology of the Academy of Sciences of the USSR). When the plants reached the stage of two pairs of true leaves (i.e., before differentiation of apices) they were placed in darkness for varying periods and then transferred back to 18-hour day conditions. The experiment included the following sets: 1) control—LD only,

FIGURE 2.1. Influence of continuous darkness on growth of hemp plants. 1 = Control plants; 2 = 1 DD; 3 = 3 DD; 4 = 5 DD.

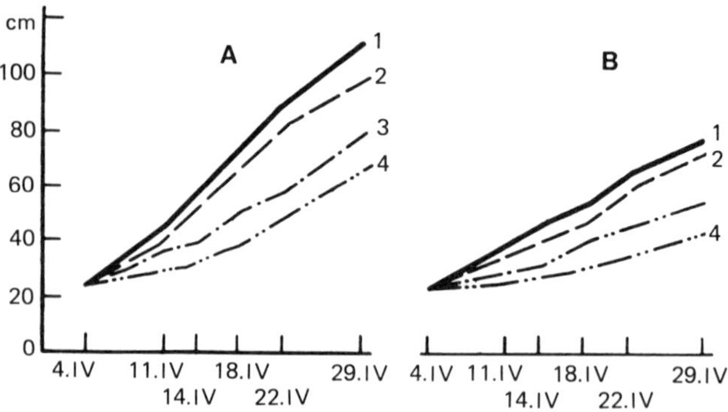

FIGURE 2.2. Influence of continuous darkness on growth of (A) male and (B) female hemp plants. 1 = Control; 2 = 1 DD + LD; 3 = 3 DD + LD; 4 = 5 DD + LD.

2) 1 dark day and then long days (1 DD + LD), 3) 3 dark days and then long days (3 DD + LD),4) 5 dark days and then long days (5 DD + LD). Each set was composed of 100 plants (50 plants in two boxes). Figures 2.1 and 2.2 clearly demonstrate that the longer the plants were kept in the dark, the more retarded was their growth. Thus, the height of male plants before harvest was 113 cm for the control and only 66 cm for the 5 DD + LD set; in females the height was, respectively, 79 and 46 cm. Continuous darkness also retarded the budding and flowering process in the male plants (Table 2.1). In the plants of the 5 DD + LD set, flowering started 8 days later than in control plants. More importantly, the sex ratio was altered under the influence of darkness: the number of female plants was increased (Table 2.1). Female individuals represented 44% of the controls, but 60% of the plants given 5 DD + LD. The data show clearly that the longer the plants are kept in the dark the more extensive is their female sex formation.

The experiment presented above thus demonstrate a direct effect of photoperiods and of constant darkness on plant sex expression.

F. Surgical Manipulations, Grafts, and Other Factors

Changes in sexual characteristics of plants can occur as a result of mechanical damage (decapitation, clipping of shoots, ablation of buds or flowers or flower clusters). This is illustrated by experiments on hemp (Pritchard 1916; McPhee 1924; Maekawa 1929; Grishko 1935; Chailakhyan 1937), asparagus and false hellebore (Levitskyi 1925), Eucommia (Bosse 1935; Kalantyr 1947), corn (Molotovskyi 1940), poplar (Starova 1969), and willow (Malutina 1973). In all these experiments, traumatic injuries inflicted on female and male plants resulted in the formation either of bisexual flowers or of flowers of the opposite sex. When plants, hemp in particular, are mechanically damaged, the morphologic changes in flower structure are identical to those induced by short-day conditions

TABLE 2.1. Influence of continuous darkness on flowering and sex expression in hemp.

Growth conditions	Beginning of bud opening (April)	Beginning of flowering (April)	Sex of plants (%)	
			Male	Female
Control (LD)	10	19	56	44
1 DD + LD	12	21	49	51
3 DD + LD	14	23	45	55
5 DD + LD	18	27	40	60

DD = day(s) in darkness; LD = long days.

(Levchenko 1937). Attempts have been made to change the sex by grafting scions from plants of the opposite sex. Although Strasburger (1900) failed to change the sex of hemp and *Mercurialis* in his grafting experiments, Grishko (1935) did obtain positive results in his reciprocal grafts of hemp plants of opposite sexes. Specifically, he observed the appearance of pistillate flowers on a male plant that had been grafted to a female plant; these flowers, however, did not yield any seeds. To understand the biology of such changes, Friedlander et al. (1977) made grafts on cucumber plants that had very different genotypes: monoecious female, and monoecious growing only on short days. It was thus discovered that the stock influenced the type of flowering on the graft: a female stock enhanced female sex expression in a monoecious graft when compared to an autograft. This influence was the strongest when the stock was young. Friedlander et al. believe that the sex-regulating substances flow from the stock to the graft (Experiment explanation in following chapters). Experiments designed to determine the influence of the graft on sex expression in the stock did not yield conclusive data. Reciprocal grafts performed on male and female *Mercurialis* plants did not result in any change of sex (Durand 1967).

Artificial regulation of the ratio of male to female flowers in hemp can be brought about by other factors. The density of sowing turned out to influence the ratio: the percentage of female plants is increased in dense sowings (e.g., Safarova 1961; Saitov 1964). This increase is probably due to diminished illumination, which was shown above to favor female sex expression. Some researchers have attempted to correlate sex expression with the quality of seed (Strasburger 1910; Sprecher 1913; Fisch 1920; Hirata 1927) or even with the location of seeds in the flower clusters (Lubich 1950). However, a detailed investigation conducted by Grishko (1935) led to the conclusion that the ratio of male to female plants did not vary significantly with respect to size, weight, color, or density of sowing of seeds (but see below).

Several attempts have been made to regulate sex expression by pollinating plants with pollen of different ages (Cieselski 1911; Lilienfeld 1921; Bessey 1918, 1933). The results obtained are contradictory. Cieselski observed that flowers pollinated with fresh pollen yielded seeds that gave rise to up to 90% male plants, but when the pollen was 12 hours old or older, the progeny was mainly female (90% to 100% of all plants). He therefore concluded that those pollen grains that carry male sex determinants have a diminished longevity. However, Lilienfeld and Bessey were unable to repeat Cieselski's results.

In conclusion, a large body of experimental work has shown that a wide variety of environmental factors and surgical manipulations result, to varying degrees, in the change of sex of dioecious plants and also of monoecious plants carrying unisexual flowers. The major determining factors studied have been mineral nutrition, atmospheric gas composition and day length. It is clear that the mechanisms that determine sex expression are very labile and that these plants are potentially bisexual.

There are two other environmental conditions that have been reported to modify the ratio of male to female flowers, but neither one has been studied in much detail. The first is drought, or water stress. In 1970, Itai and Vaadia found that the transport of cytokinin out of the roots into the shoots was decreased under water stress (Plant Physiol. 47:87–90, 1970). Since cytokinin favors femaleness (see following chapters), this should mean that dry conditions would shift the ratio toward maleness, and that is exactly what Freeman and Vitale (Botan. Gaz. 146:137–142, 1985) found with spinach plants raised in growth chambers. Here, the light fluence, temperature, day length, and humidity were all controlled. When water supply was limited, male plants flowered earlier, and set more flowers, than females. In the field, male flowers were disproportionately abundant, and the ratio was further modified in dry conditions.

The second environmental influence is the crowding or sparseness of plants in the field. Doust, O'Brien, and Doust (Am. J. Botan. 74:40–46, 1987) recently reported that in the wild pink, *Silene alba* of the Caryophyllaceae crowding in the field favored femaleness. The female plants had 25% less leaves, but 10% more roots (by dry weight) than the males. The male:female ratio was about 1:2, but was modified by high density. This may well be related to the water stress influence above. [Ed.].

3
Hormonal Regulation of Sex Expression and Age-Related Changes

A. Influence of Phytohormones and Growth Inhibitors on Sex Expression in Whole Plants

In the course of working out the effects of the phytohormones (auxins, gibberellins, abscisic acid, ethylene), researchers have studied their effects on growth and development and also on sex expression. By analogy with animals, in which the importance of hormones in sex regulation has long been established, it was expected that phytohormones would play a significant role in plant sex regulation. In fact, a role of growth regulators in modifying sex expression, both in dioecious plants and in monoecious plants with unisexual flowers, has now been demonstrated in a number of studies (Heslop-Harrison 1957; Vince-Prue 1975; Frankel and Galun 1977; Sidorskyi 1978; Minina and Larionova 1979).

Auxins

It goes without saying that auxins are involved in a large number of physiological and biochemical processes and that they act as endogenous growth regulators participating in the division, elongation, and differentiation of cells of many types (Kholodnyi 1939; Turetskaya 1961; Polevoy 1970; Kefeli 1973; Gamburg 1976).[1] The first effect of auxins on sex expression was shown in experiments on cucumber carried out by Laibach and Kribben (1950, 1951). After treatment of cucumber leaves with a lanolin paste containing naphthylacetic acid (NAA), Laibach and Kribben observed a significant increase in the number of female flowers. The same result was subsequently obtained with many representatives of the gourd family (Heslop-Harrison 1959; Pykhtina 1971; Lvova 1973; Bisaria 1974; Corley 1976). Enhancement of female sex expression by NAA treatment was observed in hemp as well (Vergely et al. 1967). Lvova (1973) considered that

[1]A more complete and recent account is given by K. V. Thimann (Hormone Action in the Whole Life of Plants. Amherst, Massachusetts: University of Massachusetts Press, 1977 [Ed.].

NAA could induce the change of sex only if the flowers were treated at an early stage of organogenesis (from appearance of floral meristems to the formation of sepals in embryonic flowers). NAA treatment of plants at later stages either fails to influence sex expression or induces such intensive proliferation and differentation of carpels that microsporogenesis and stamen development are interfered with.[1] A decrease in the viability of pollen as a result of NAA treatment was observed not only in *Cucumis sativa* L. but also in experiments on the hermaphroditic LD plant *Silene pendula* (J. Heslop-Harrison and Y. Heslop-Harrison 1958a). Interestingly enough, the effect of NAA in favoring female sex expression is enhanced if melon plants are treated with the substance under SD conditions (Kaushik and Bisaria 1974).

Other active synthetic analogs of auxins include the salts of NAA and 2,4-dichlorophenoxyacetic acid (2,4-D). Treatment of cucumber plants with the sodium salt of NAA resulted in a 17-fold increase in the number of pistillate flowers on the main stem and in a two- to threefold decrease in the number of staminate flowers. Studying the effects of the natural auxin, indoleacetic acid (IAA), Heslop-Harrison discovered that it led to the formation of hermaphroditic flowers on male hemp plants, but that it did not affect sexuality in mercury or evening campion (Heslop-Harrison 1964). With several members of the gourd family also, IAA treatment led to an increase in the number of female flowers (Galun 1959a; Heslop-Harrison 1959; Sedlovskyi 1972), and there were similar effects in corn (Molotkovskyi 1957; Heslop-Harrison 1961; Sladky 1966, 1969), and Begonia (Heide 1969). Galun et al. (1962) cultivated cucumber flower buds on a nutrient medium in vitro; addition of IAA transformed staminate flowers into pistillate flowers. It should be noted that the male cucumber tissue proved to be more sensitive to IAA in the medium than the female tissue. Injection of IAA into the stem of henbane (*Hyoscyamus niger*) retarded flowering but resulted in the proliferation of gynaecium and calyx in the buds that did subsequently develop (Resende 1953). Based on their own studies on hemp (Heslop-Harrison 1956; J. and Y. Heslop-Harrison 1957) and on other researchers' work on the *Cucurbitaceae* (Laibach and Kribben 1950a,b; Nitsch et al. 1952; Wittwer and Hillyer 1954), J. and Y. Heslop-Harrison came to the conclusion that high levels of auxins in the tissues adjacent to differentiating flower primordia favored pistil development; alternatively, low levels of auxins favored stamen development. This conclusion was supported by the observation that an artificial increase in auxins during differentiation led to the formation of female flowers at exactly the same loci as would be normally occupied by male flowers. It was also found that enzymatic oxidation of IAA (by reaction with IAA-oxidase, peroxidase, or catalase) is much more rapid in the embryonic flowers of female cucumber lines than in those of male cucumber lines. Because 2,3,5-triiodobenzoic acid (TIBA) delays the polar movement of auxins in plant tissues, its application should lead

---

[1]This general conclusion is confirmed experimentally in Section C of this chapter [Ed.].

to a build-up of auxins in specific sectors. This explains the auxin-like effect of TIBA on plant sex expression (J and Y. Heslop-Harrison 1957; Frankel and Galun 1977). Alternatively, in the experiments of Sladky (1974), TIBA acted as an anti-auxin and induced extensive formation of staminate catkins in the walnut. Galun (1959b) determined the auxin activities of the organs of cucumber plants, but initially failed to find differences between male and female cucumber lines. However, subsequent experiments on male and hermaphroditic cucumber lines demonstrated that a higher auxin activity is characteristic for hermaphroditic plants (Galun et al. 1965). Rudich et al. (1972a) showed that the tissues of female cucumber plants contain higher levels of auxins and lower levels of gibberellins than male tissues. The specific role of auxin in sex regulation and sex expression has also been emphasized by other researchers (Fujii 1972).

Thus, the majority of experiments indicate that auxins increase the number of flowers of the female sex type; however, the mode of action of auxins in this process remains unclear. In this respect, Heslop-Harrison (1963) believes that auxins act as a triggering mechanism. Above a certain threshold concentration in the apex, they are considered to activate a genetic system that previously was dormant. Sidorskyi (1978) proposed that auxins influence plant sex formation by regulating carbohydrate metabolism. This hypothesis is based on the fact that in the melon, hermaphroditic flower buds have a higher content of carbohydrates and auxins than male flower buds (Randhawa and Singh 1973). Moreover, an increase in the number of male flowers in cucumber plants is accompanied by a decrease in the reducing sugar content (Sidorskyi and Sidorskaya 1973; Sidorskyi et al. 1973).

Gibberellins

Numerous studies indicate that with respect to plant sex expression, gibberellins have the opposite effect to that of auxins. In the majority of cases gibberellins induce male sex expression. Unfortunately, gibberellin-induced changes in sex have only been well defined for a relatively small number of species (hemp and cucumber, among others).

The earliest studies of the effect of gibberellins on plant sex expression were conducted on hemp. According to Atal (1959), treatment of genetically female plants with gibberellins resulted in the formation of intersexes and of male flowers with normal pollen. Atal believes that the transformation of sex by gibberellin occurs during flower differentiation. Zhukov and Sazhko (1963) observed that in monoecious hemp, gibberellin treatment increases the number of feminized male plants. At the same time, female plants treated with gibberellin produced male flowers (Zhukov and Sazhko 1963; Gorshkov and Sazhko 1964). In Köhler's experiments (1964), gibberellin induced the elongation of flower clusters and the formation of male flowers on female plants. However, not all the plants responded in this manner; the plants that did were presumably predisposed to the formation of male flower primordia.

These experiments on hemp have shown that gibberellin treatment resulted in a number of morphological changes: a significant elongation of flower clusters, a decrease in the size of male flowers and in the quantity of pollen, and formation of male flowers on female flower clusters 8 to 12 days after they have formed seeds (Zhukov and Sazhko 1961; Zhukov et al. 1963). Another effect of gibberellin on hemp was that treatment of a crop of dioecious hemp with GA at the age of 28 to 35 days resulted in the appearance of a large number of monoecious plants. Mohan Ram and Jaiswal (1972) and Jaiswal and Mohan Ram (1974) showed that gibberellin induces the formation of male flowers on female plants and generally increases their number on male plants. Because cycloheximide attenuates the plants' response to gibberellin (Chapter 6), the authors suggested that cycloheximide inhibited one of the enzymatic reactions that lead to the formation of male flowers. A parallel result was reported with hops (*Humulus lupulus*) in which gibberellin treatment led to a reduction in the number of pistillate flowers (Zattler and Chrometzka 1968).

Gibberellin triggers important changes in the formation of the sexual characteristics of the *Cucurbitaceae*. It was Wittwer and Bukovac (1958) who first discovered that gibberellin treatment of cucumber plants resulted in an increase in the size of staminate flowers. Galun (1959b) worked with female heterozygous plants that carry exclusively female flowers on the lateral shoots and on the upper nodes of the main stem. Repeated treatments of these plants with GA resulted in modification into normal monoecious plants, that is, plants bearing both male and female flowers. Histologic and radiographic experiments conducted by Fuchs et al. (1977) showed that the treatment of monoecious and genetically female cucumber plants with a solution of gibberellins $A_4 + A_7$ at 15 μg/mL inhibited the development of pistillate flowers and stimulated the development of male flowers. The authors believe that these gibberellin-induced staminate flowers do not arise from the transformation of pistillate or bisexual flowers toward maleness, but are in fact additional flowers that never appear in normally developing cucumber plants. Such changes in the sex of *Cucumis sativus* are most effectively induced by the gibberellins A_7, A_4, A_2, A_3 (Wittwer and Bukovac 1962) and A_{13} (Clark and Kenney 1969). Although gibberellin A_3 is less effective than A_7 and A_4 (Pike and Peterson 1969), the former is more easily accessible, so that the majority of experiments have been conducted using gibberellic acid, GA_3. Moreover, A_3 turned out to be the most effective gibberellin for inducing male sex formation in other species of *Cucurbitaceae*, such as *Luffa acutangula* (Krishnamoorthy 1972). Vlasenko (1973) induced male sex formation in cucumber by treating the plants with gibberellin and growing them in 10-hour morning light. Lvova (1973) noticed that gibberellin affects sex expression in cucumber if the plants are treated at an early stage of flower organogenesis. Sidorskyi (1972) observed that the gibberellin-induced change in the sex of flowers is independent of the sex type of the plant over a large portion of the shoot. This effect of gibberellins, namely stimulation of the formation of male flowers and inhibition of the formation of female flowers, has been noted by many investigators with different species (Peterson and Anhdler 1960; Rudich et al. 1972; Rudich and Halevy

1974; Nozzolillo 1972).[1] Perhaps the most characteristic results have been obtained with other members of the gourd family, such as *Cucumis melo* (Brantley and Warren 1960), and *Luffa acutangula* (Bose and Nitsch 1970; Krishnamoorthy 1972).

Thus, GA treatment of watermelon plants (*Citrullus lanatus* [Thunb. Munsf.]) also led to an increase in the number of male flowers (Bhandari and Sen 1973). Kalyagin (1973) demonstrated in his experiments on melon that gibberellin induces male sex formation in plants that normally would strongly express female sexual characteristics, and that gibberellin does not affect the time of flowering of staminate flowers, but does delay the opening of pistillate flowers.

Kaushik and Bisaria (1974) studied the combined effect of gibberellins and day length by spraying melon plants at the two-leaf stage and growing the plants under various conditions. The following results were obtained: 1) uninterrupted daylight + GA led to an increase in male sex expression; 2) GA on short days did not significantly affect sex expression; but 3) GA on long days led to the formation of a maximal number of male flowers and a minimal number of female flowers.[2]

Pharis and co-workers (Pharis and Morf 1970; Pharis et al. 1970, 1974, 1975) have conducted detailed studies on conifers. They demonstrated that the formation of female cones requires higher concentrations of GA (2500 mg/L) than the induction and formation of male primordia (250 mg/L). The induction by gibberellin can be modulated by other factors, such as day length and temperature. For instance, in order to increase the number of male strobili on LD, low concentrations of GA (75–250 mg/L) would suffice; on SD, however, high concentrations of GA (250–500 mg/L) are required. As shown in the experiments of Pharis and Owens (1966; Owens and Pharis 1967), GA treatment of Arizona cypress leads to an increase in the number of mitoses in the subapical meristem of the apical cone; male primordia are established 18 to 22 days after GA treatment.

Gibberellin treatment (at a concentration of 50 mg/L) of *Bryophyllum* and *Hyoscyamus niger* results in the transformation of female flowers into hermaphroditic ones. The opposite changes in sex were observed when higher concentrations of GA were used (Resende and Viana 1959, see below). GA treatment

---

[1]Another effect, but in the opposite sense, occurs when corn plants are treated with GA in the first month after sowing, for this leads to feminization of the tassel. Five corn varieties showed this effect (Nickerson N.H.: Ann. M. Bot. Garden 46:17–39, 1959). Only when GA was applied later did it promote maleness [Ed.].

[2]The formation of GA, as initially observed in cultures of *Gibberella fujikuroi* but also generally in higher plants, is antagonized by chlorocholine chloride (CCC). This relation was made use of by Wang Ben-li and Tsao Tsung-hsung (Symposium on Plant Tissue Culture, Peking, pp. 511–516, 1978), who cultured sterile shoot apices of cucumber on media containing GA and CCC in various ratios. With CCC alone, about 2.5 times as many female as male flowers were formed. GA alone inhibited all female flowers. Increasing the CCC in presence of constant GA (20 mM) led to an inverse log-linear relation between CCC concentration and number of female flowers. The balance was struck at about 4 times (on a molar basis) as much CCC as GA [Ed.].

of corn also leads to the formation of hermaphroditic flowers (Sladky 1969, 1971; Krishnamoorthy and Talukdar 1976). In mercury (*Mercurialis*), gibberellin induces the formation of male flowers in internodes of female plants (Champault 1973). In his experiments on walnut (*Juglans regia*), Sladky (1974) showed that gibberellin stimulates the differentiation of flowers and leads to an increase in the number of staminate catkins.[1]

On the other hand, in a number of cases, gibberellin stimulates female sex expression. This was observed by Hashizume (1959a,b, 1960, 1961) in his experiments on *Cryptomeria japonica* and false cypress (*Chamaecyparis obtusa* and *Chamaecyparis lawsoniana*). GA treatment immediately after the formation of the microsporangium leads to the transformation of male cones into female cones. These female cones develop normally and yield viable seed. Witsch (1961) noted that spraying cocklebur plants with gibberellin under both short-day and long-day (LD) conditions increases the number of female flower clusters. Gibberellin-induced female sex expression and an increase in the number of female flowers were also observed in castor beans grown on LD by Shiffris (1961), who concluded that gibberellin exerts the same influence on sex formation in castor bean as do LD conditions. Gibberellin also induces female sex formation in spinach (*Spinacea oleracea*). Interestingly, this effect is observed only if male spinach plants are grown on LD; no change of sex occurs on SD (Chulafich 1978). This result is in contrast to the many experiments described above in which it has been shown that gibberellin induces male sex formation in numerous LD plant species (Frankel and Galun 1977). Many hypotheses have been put forward to explain these effects of gibberellin on plant sex expression. For instance, Takasi and Hideo (1964) proposed that the presence of auxins and gibberellins in the apical buds results in an influx of specific substances that are formed in the leaves under the experimental conditions. It is these substances, they feel, that are responsible for sex differentiation, and they are probably the true plant sex hormones. The presence of auxins and gibberellins in the apical buds was not by itself considered to be a factor determining the sex of different varieties of cucumber.[2]

All in all, the effects of gibberellins on sex expression have been reasonably well studied. Gibberellins are responsible for clear changes in total sex expression in only a relatively small number of plants (e.g., hemp, buckwheat, and cucumber). In the majority of other cases, gibberellins shift sex differentiation towards the male type. This observation is in good agreement with the finding that the biological activity of endogenous gibberellins is higher in male hemp plants than in females (Khrianin 1975), and that the content of gibberellins is sig-

---

[1]Still more species in which GA promotes maleness are cited by Pharis R.P., King R.W.: Ann. Rev. Plant Physiol. 36:517–568, 1985 [Ed.].

[2]Bruinsma (Acta Horticulturae 31:81–87, 1973) points out that gibberellins often inhibit flower development and that the ovary and style are the last parts of the flower to be initiated; hence they suffer the greatest delay in development under the influence of gibberellin, or may not be initiated at all. With this straightforward explanation, one need not invoke additional unknown substances [Ed.].

nificantly higher in male lines of various species of the gourd family than female lines (Atsmon 1968; Atsmon et al. 1968; Bhattacharya and Tokumasu (Engei Gakkai Zaschi 39:224–231, 1970); Hayashi et al. 1971; Hemphill et al. 1972). However, it is clear that the effects of gibberellins on sex expression vary according to the age of the plants, the competence of the apex, and with the time and the type of treatment and the concentration used (Khrianin and Milyaeva 1977; Frankel and Galun 1977; Minina and Larionova 1979).

Cytokinins

In contrast to the extensive work with gibberellins and auxins, little is known about the effects of cytokinins on sex expression. The difficulty in studying this group of phytohormones arises from the fact that, unless cytokinins are introduced into the roots, their range of action is usually limited to the site of application (Kulaeva 1973). Therefore, many investigators have applied cytokinins directly to the flowering buds or shoots. The influence of cytokinins on sexuality has been best studied in different varieties of grape. Negi and Olmo (1966) maintained flowering shoots of a male grape clone in a solution of 6-BAP[1] for a period of three weeks. They observed a complete transformation of male flowers into hermaphroditic ones. None of the other growth regulators tested (IAA, GA, TIBA) had any effect on sex expression in this clone. It should be noted that this particular male clone, a representative of *Vitis sylvestris*, belongs to a group of grape plants that bear very few hermaphroditic flowers under natural conditions, because of interference with the formation of the ovary, and the resulting underdevelopment of the pistil. However, cytokinin restores normal pistil development and thus produces functionally hermaphroditic flowers in this clone (Negi and Olmo 1972). A cytokinin-specific increase in the number of functionally bisexual flowers in male grape clones has been observed in other experiments (Moore 1970; Hashizume and Iizuka 1971; Mullins 1980). Moore (1970) believed that the flower structure of grape plants is not dependent solely on the genes themselves, but is also under the control of gene modifiers, the influence of which may change when the plants are treated with cytokinins. Hashizume and Iizuka (1971) found that not only the synthetic cytokinins but also the natural ones, zeatin in particular, induce the formation of functionally normal flowers on male grape shoots.

Cytokinin-induced changes in the type of sex expression have been noted in other plant species. Thus, spraying *Mercurialis* with kinetin led to the transformation of male flowers into female ones in 45% of the treated plants (Durand 1967). Interestingly, when kinetin was applied in conjunction with GA or IAA, the effect on sex expression was nil. The cultivation of male *Mercurialis* stem cuttings on a cytokinin-rich medium also leads to the formation of pistillate flowers in male flower clusters (Champault 1970, 1973). In vitro culture experiments conducted on *Cleome iberidella* showed that zeatin stimulates the differentiation and growth of the pistil (Jong and Bruinsma 1974a,b). Cytokinin also affects sex

[1]Benzylaminopurine, a synthetic cytokinin of high activity [Ed.].

expression in Bryophyllum; female sex formation is favored in hermaphroditic flowers (Catarino 1964). Unfortunately, data on the effect of cytokinins on sex expression in *Cucurbitaceae* are practically nonexistant. Although the results obtained are rather limited, it is nonetheless clear that cytokinins, just like other phytohormones, play an important role in plant sexual differentiation.

Abscisic Acid and Retardants

In addition to stimulatory hormones, sex expression is also regulated by inhibitory hormones. The best studied inhibitory phytohormone is abscisic acid (ABA), which is undoubtedly a natural regulator of plant growth and development (Ohkuma et al. 1963; Paseshnichenko 1978). With respect to the internal regulation of cellular metabolism, ABA has been shown to act in conjunction with, or sometimes as an antagonist of, gibberellins (Addicott and Lyon 1969), auxins (Kefeli 1974; Gamburg 1976), and cytokinins (Kulaeva 1973; Kravyazh et al. 1977). In cucumber, for instance, abscisic acid participates in sex regulation by counteracting the effect of gibberellin (Rudich and Halevy 1974; Rudich et al. 1972a); the experiments showed that the treatment of cucumber plants with a solution of abscisic acid stimulates the formation of female flowers. In their experiments on pumpkin, Abdel-Gaward and Ketellaper (1969) showed that ABA is similar to SD conditions in that both enhance the initiation of female flowers. ABA by itself does not affect sex expression in hemp; however it does eliminate the tendency to male sex expression caused by gibberellin (Mohan Ram and Jaiswal 1972). According to Engelbrecht (1973), male hemp flower clusters have a characteristically high ABA content, while female flower clusters contain mainly cytokinins. Engelbrecht therefore proposed that the balance between these two growth regulators could be a major factor in determining phenotypic differences between male and female plants. This view suggests that ABA is one of the regular components of the hormonal system that regulates sex expression in plants.

In addition to natural inhibitors, sexual differentiation in plants is affected by a number of synthetic inhibitors, retardants, and other physiologically active substances (Muromtsev and Khrianin 1974; Frankel and Galun 1977; Sidorskyi 1978). Chlorocholinechloride (CCC) is a synthetic growth inhibitor used to prevent the lodging of cereals (Tolbert 1960a,b; Prusakova et al. 1970; Deeva 1980). With respect to the growth and development of the majority of plants, this retardant has an effect opposite to that of gibberellins (Muromtsev and Khrianin 1974; Chailakhyan and Nekrasova 1976). For instance, treatment of many *Cucurbitaceae* (cucumber, melon, and pumpkin, among others) with a solution of CCC leads to a significant increase in the number of pistillate flowers and a decrease in the number of staminate flowers (Tekhanovich 1970; Mishra and Pradhan 1970; Ghosh and Bose 1970). However, in the case of spinach and hemp (Chailakhyan et al. 1969), CCC treatments do not alter the ratios of male to female plants; nor is sexual differentiation in the oil palm affected by this retardant (Corley 1976). Champault (1973) determined that CCC does not affect sex expression in *Mercurialis*; however, when used in conjunction with GA, CCC suppresses the effect of the GA. Similar experiments conducted on another variety of hemp (Muromtsev and Khrianin 1974; Khrianin 1978) showed that CCC enhances female sex expression and reduces the male sex expression caused by GA, yet total suppression does not take place. This effect of CCC, i.e., the increase in the number of female flowers, can be partially explained by the well-known inhibition of GA synthesis by this retardant (Dennis et al. 1965; Barnes et al. 1969; Cross and Myers 1969) and by its stimulation of the synthesis of growth inhibitors (Ivanova 1971). As suggested by Rudich and Halevy (1974), a high ratio of inhibitors to gibberellins could favor female sex formation.

The numerous studies of the effects of phytohormones and retardants on sex expression in various plant species have made it possible to assign specific roles to these substances. It thus appears that in general, gibberellic acids are hormones that enhance maleness whereas auxins and, in certain cases, cytokinins and ABA are phytohormones that enhance femaleness. However, the body of experimental evidence includes many exceptions and even some contradictory results (discussed above). It need not surprise us, therefore, that several different models have been created to explain the modes of action of growth regulators on sex expression in plants. In some cases, the predominant role in the regulation of sex expression was assigned to the endogenous level of auxins (Heslop-Harrison 1957, 1963); in other cases the levels of gibberellins (Atsmon 1968) or of ethylene (Loy 1971; Byers et al. 1972) were considered to be critical. Some authors believe that plant sex formation is controlled by the balance between phytohormones, and others by the balance between phytohormones and inhibitors. Specifically, the following parameters have been considered to be important: the ratio of auxins to gibberellins (Kutuzova 1969; Pharis and Ross 1976), of cytokinins to abscisic acid (Engelbrecht 1973), of gibberellins to abscisic acid (Rudich and Halevy 1974), and that of phytohormones to natural inhibitors (Rudich et al. 1972; Minina 1973).

All of these divergent views on the regulatory mechanism of sex expression in plants arose mainly because the effects of phytohormones and inhibitors were tested in different species of plants, at different ages, and at different stages of development. Moreover, in the majority of cases, the introduction of phytohormones and inhibitors was achieved by merely spraying the aerial parts of the plants, without taking into account the roots, or the levels and biological activities of endogenous phytohormones and inhibitors.

Because of the limitations of the studies that had been conducted, it appeared probable that further investigations would bring more insight into the problem of hormonal regulation of sex expression in plants.

B. Phytohormone Treatment of Seeds Before Sowing

The effects of treating seeds with phytohormones before sowing have been studied in a number of different plants. Molotkovskyi (1957) carried out an experiment in which corn seeds were soaked in a 0.001% solution of 2,4-D for a period of 10 days. The formation of ears in the (male) tassels was observed in 15% of all treated plants. An increase in the number of female flowers of *Luffa acutangula* was obtained when the seeds were initially soaked in a solution of benzyladenine (Bose and Nitsch 1970). Herich (1960, 1961) achieved a small increase in the number of female hemp plants by soaking the seeds in a solution of gibberellin. However, Davidyan and Kutuzova (1970) have failed to reproduce Herich's results.

The experiments presented below were conducted on hemp (*cv.* US-6) and spinach (*cv.* Victoria). The seeds were treated with the following substances: IAA, GA, 6-benzylaminopurine (6-BAP), and ABA. Hemp and spinach seeds were soaked in a solution of a given growth regulator for 24 hours. The following concentrations were used: 1) GA: 25 and 50 mg/L; 2) IAA: 25 and 50 mg/L; 3) 6-BAP: 25 and 50 mg/L; and 4) ABA: 10 and 20 mg/L. Control seeds were soaked in water under the same conditions. After this treatment the seeds were sown in soil both in the field and in the greenhouse.

The field experiment was conducted under the usual culture conditions, with natural day length (locations: Young Naturalists' Station of Pemz). Each plot contained 90 hemp plants and 70 spinach plants. The experiment showed that soaking seeds in GA led to a *decrease* in the number of female plants in both species. This decrease was matched by an increase in the number of monoecious hemp plants and spinach intersexes. IAA and 6-BAP treatments did not alter the sex ratios in hemp, but both of them caused significant decreases in the relative number of male spinach plants. Treatment with ABA resulted in a decrease in the number of male hemp plants owing to the appearance of intersexes; no changes in sex ratios were observed in spinach.

In the greenhouse experiment, artificial lighting (xenon lamps, 14,000 lux, temperature 25 °C) was used at the Timiryazev Plant Physiology Institute. Plants were grown in boxes containing soil, 65–70 hemp plants or 45–50 spinach plants per box. The young plantlets were transferred to conditions that favor flowering, i.e., 8-hour days for hemp and 18-hour days for spinach. The experiment demonstrated that GA significantly stimulates the growth in height of hemp plants; any stimulation by IAA was too slight to be significant. 6-BAP and ABA, when used at low concentrations, did not affect the growth of the stem; higher concentrations of these substances did inhibit the growth in height of both species.

In the greenhouse, as in the field, GA treatment of hemp seeds led to a decrease in the number of female individuals and to the appearance of monoecious plants. On the other hand, 6-BAP or IAA treatments resulted in a decrease in the number of male plants owing to the appearance of intersexes. A similar effect, although less pronounced, was observed with hemp seeds treated with ABA. In spinach, GA stimulated the growth in height of the plants by 65%, compared with the control. IAA proved to be less of a growth stimulator. In contrast, 6-BAP (50 mg/L) and ABA (20 mg/L) acted as growth inhibitors. The treatment of spinach seeds with GA resulted in a small decrease in the number of female plants and in the appearance of intersexes. IAA also slightly increased the number of female individuals, whereas 6-BAP and ABA treatments led to a significant increase in the number of male spinach plants (Figure 3.1).

FIGURE 3.1. Male (A) and female (B) spinach plants grown from seeds treated with different growth regulators. 1 = Control, treated with water; 2 = GA; 3 = IAA; 4 = ABA (10 mg/L); 5 = ABA (20 mg/L); 6 = 6-BAP (25 mg/L); 7 = 6-BAP (50 mg/L). Greenhouse experiment.

The greenhouse experiment thus leads to the following conclusions.

1. The growth of both plant species is stimulated by GA, and, to a lesser extent, by IAA. Treatment of seeds with 6-BAP or ABA at high concentrations results in growth inhibition.

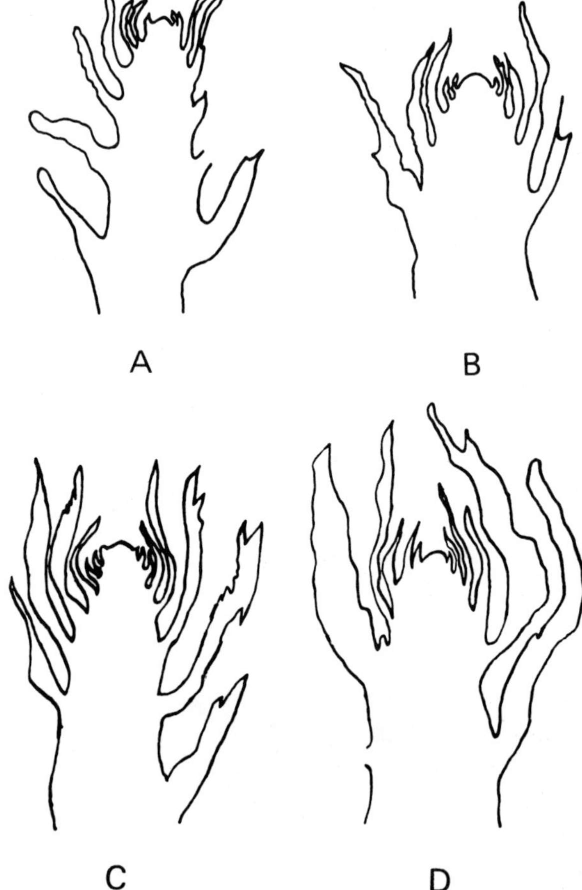

FIGURE 3.2. Schematic representation of longitudinal sections of stem apices of male (A, C, E, G) and female (B, D, F, H) hemp plants. A,B = the stage of three leaf-pairs; C,D = 22 hours later; E,F = 40 hours later; G,H = 14 days later.

2. None of the substances tested appears to stimulate flowering. Treatment of seeds with 6-BAP or ABA even delays the flowering of hemp and spinach plants.

These data confirm the results that had been previously obtained by spraying plants with aqueous solutions of phytohormones and inhibitors. Hemp and spinach plants display an enhancement of male sex expression by gibberellin and an enhancement of female sex expression by IAA. The effects of cytokinin (6-BAP) and ABA are less clear, since overall, neither of these compounds caused substantial changes in the sex ratios of the two plant species.

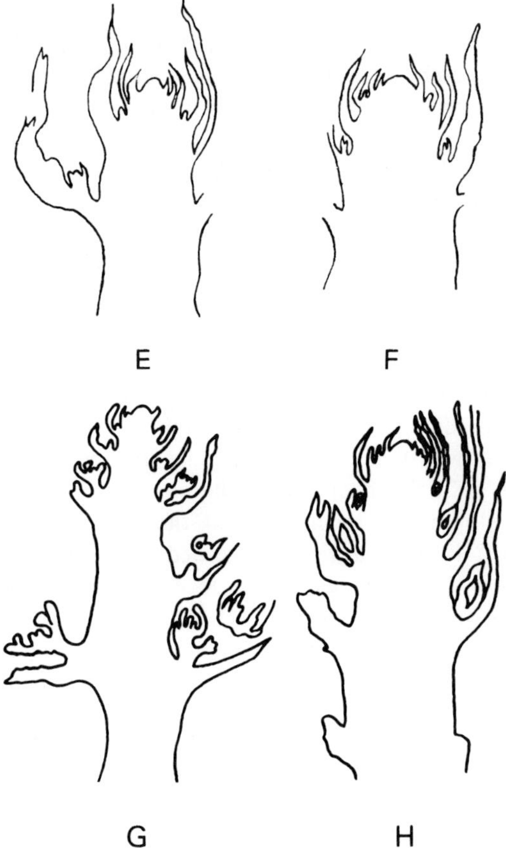

FIGURE 3.2. (*Continued*).

C. Sex Expression and the Age of the Plant

An important consideration in studying the regulation of sex expression by external physical or chemical factors is the time at which the plants are treated. To deal with this variable, a basic question needs to be answered: when, i.e., at what developmental stage of plants in general, is sexuality initiated? This question is particularly critical for dioecious plants and monoecious plants with unisexual flowers. It appeared that a reasonable answer to this question could be provided by studying the dynamics of development of the apex with special reference to age-related changes. This, in turn, could help locate that critical period during which flower primordia and their sexual differentiation are established, i.e., when the sex of the plant is determined.

Anatomical studies of stem apices of hemp have revealed the timing of the major age-related changes in development (Khrianin and Milyaeva 1977). Initially the apex of the embryonic hemp plant is practically flat. Two to three days after sprouting, the center of the apex rises and the apex thus acquires a slightly convex shape. With respect to sexual differentiation, no morphological distinctions can yet be made. The apices of hemp plants remain in a vegetative state until after the formation of three pairs of leaves. At this point, the apices enter a prefloral stage, in which the apical cone becomes dome-shaped and reaches its maximal dimensions. Although the generative primordia are still absent, the structure of the apex makes it possible to distinguish, even at this early stage, between male and female plants. The apices of male plants are narrow and elongated (Fig. 3.2A), whereas those of female plants are wide and short (Figure 3.2B). This trend is maintained at later stages, the third internode of male plants remaining more elongated than that of the females (Khrianin 1964). Recent experiments of Lacombe (1980) showed that in the male plant the third internode at a given growth stage is up to 25% longer than in the female.

Thus, in the development of the hemp plant, the stage of three leaf-pairs is critical as far as sexual differentiation is concerned. Interestingly, it is at this stage that one observes a sharp increase in the levels of auxins and particularly of gibberellin-like substances (Khrianin 1974). At 22 hours after the beginning of the three-leaf pair stage no significant changes in the apices are detectable (Figure 3.2C,D), but 40 hours later, generative primordia appear in both male (Figure 3.2E) and female (Figure 3.2F) plants; the differentiation of flower-buds is initiated. Clear contrasts between the differentiation of male and female flowers appear subsequently. The structures of the apices of male and female plants after 14 days of growth are shown in Figure 3.2. Both male flowers (Figure 3.2G) and female flowers with their ovules (Figure 3.2H) are clearly differentiated. Thus, the study shows that generative primordia appear at the end of the

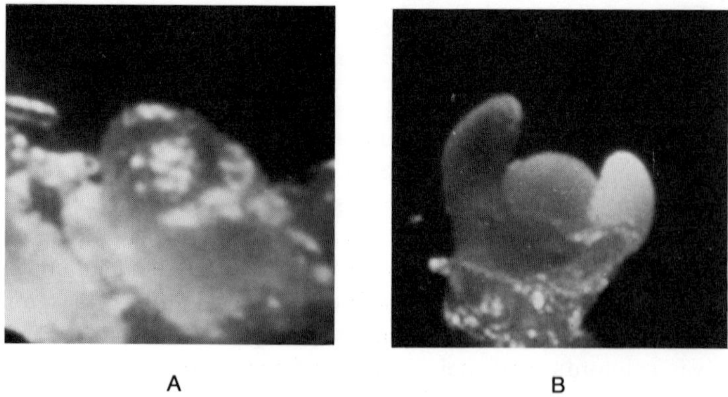

FIGURE 3.3. Vegetative apices of (A) male and (B) female spinach plants.

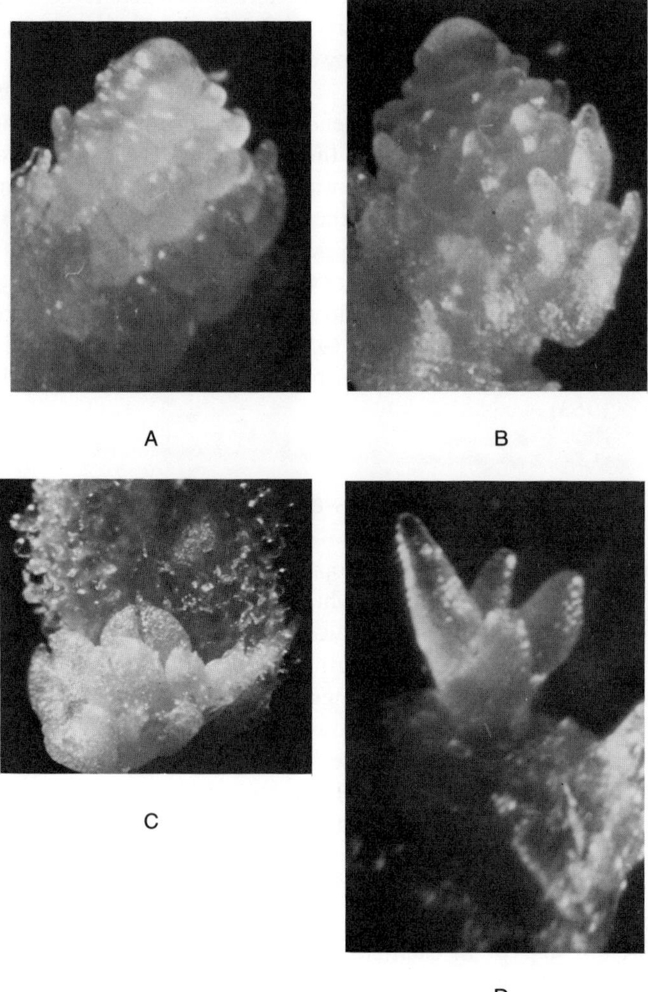

FIGURE 3.4. Apices of spinach plants carrying differentiated male (A) and female (B) flowers; anthers (C); female flower (D).

three leaf-pair stage of hemp development, and their sexual differentiation begins only a few days later.

Experiments with spinach grown in LD in boxes of soil showed that the morphological changes of stem apices that occur in the course of development are similar to those observed in hemp. Specifically, at the one-leaf stage (10 days after sowing), generative primordia have not yet appeared, but the apices already differ in their overall structure (Figure 3.3). At the beginning of the formation of the third leaf, the apices leave the vegetative state and enter a reproductive stage;

this shift is shown by the appearance of undifferentiated generative primordia. At the five-leaf stage, some apices carry well-differentiated male flowers with anthers, while others carry female flowers (Figure 3.4).

Thus, monitoring the development of apices has clearly shown that in order to regulate sex expression with exogenous phytohormones and inhibitors, the substances must be introduced into the plants at the earliest developmental stage possible, i.e., before the differentiation of flower primordia. Specifically, hemp plants should be treated before the three leaf-pair stage and spinach plants before the formation of the third leaf. Moreover, it is not sufficient to conduct phenological observations: the state of the apex should be monitored as well. In corn, the differentiation of generative primordia in the apex takes place at the three-leaf stage (Kuperman et al. 1955). In cucumber, the differentiation of the apices and the formation of flower primordia occurs already at the one-leaf stage (Lvova 1963).

D. Effects of Phytohormones at Early Developmental Stages

It was shown above that in the development of hemp, spinach, cucumber, and corn there is a critical period during which flower primordia are established and undergo differentiation. This finding was followed by a number of solution culture experiments in which phytohormones were introduced into the plants through the root system, using seedlings at early developmental stages.

The first experiment was conducted on hemp (*cv.* US-6). Plants were grown in solution cultures in the greenhouse of the Plant Physiology Institute of the Academy of Sciences of the USSR. Hemp seeds were germinated on wet filter paper for three days, at 25°C, in the dark. The seedlings that had roots of comparable size were then transferred to 1- or 3-L vessels filled with water. After 2 days, the plantlets were transferred to test solutions in the following groups: 1) controls—water, 2) GA (25 mg/L), 3) IAA (15 mg/L), 4) 6-BAP (15 mg/L), and 5) ABA (10 mg/L). In each case, 70 plants were treated. After 24 hours the plantlets were transferred to a dilute Knop's solution (0.1 × normal concentration), then 2 days later to a half-strength Knop's solution, and after 2 additional days to full-strength Knop's solution that was subsequently aerated on a daily basis. The plants were grown under short, 8-hour day conditions that lead to full sex expression.

This experiment, like those described, showed that growth in height is considerably stimulated by GA and inhibited by 6-BAP. Plants treated with IAA or ABA had a growth rate comparable to that of the controls (Figure 3.5, Table 3.1). It should be noted also that the leaves of the plants treated with 6-BAP were dark green throughout their growth, whereas the leaves of the plants treated with GA were light green. By 15 days after the beginning of the experiment, the roots of the 6-BAP-treated plants had developed tumor-like structures. Inspection of cross-sections under the microscope revealed that these structures were due to a proliferation of cell tissue—a direct result of the phytohormone treatment. Bud

FIGURE 3.5. Growth and sexuality in hemp after introduction of phytohormones into the roots. 1 = Control; 2 = GA; 3 = BAP; 4 = IAA; 5 = ABA.

development in GA-treated plants occurred 4 days earlier than in the controls; IAA, on the other hand, retarded budding. In all the other treated plants, budding took place simultaneously with the control plants. Interestingly, development of buds occurred after the formation of three-leaf pairs in the case of all the treated plants, with the exception of those given 6-BAP; these started bud growth after the second pair of leaves had formed.

TABLE 3.1. Effect of phytohormones on growth and development of hemp (introduction through the roots).

Treatment	Height of plants (cm)[1]			Beginning of bud development (April)
	Males	Females	Intersexes	
Control	31.0	16.0	25.0	18
GA	46.0	39.0	41.0	14
6-BAP	0	6.5	8.0	18
IAA	0	17.0	26.5	22
ABA	27.0	14.5	22.5	18

[1]Average of 10 measurements.

TABLE 3.2. Effect of phytohormones on sex expression in hemp (introduction through the roots).

	No. of plants (%)		
Treatment	Males	Females	Intersexes
Control	28.6	37.0	34.4
GA	84.2	6.5	9.3
6-BAP	0	47.2	52.8
IAA	0	40.0	60.0
ABA	19.6	38.8	41.6

The effects on sex expression of introducing growth regulators through the roots of the seedlings are shown in Table 3.2. In the case of GA, the majority of seedlings developed into male plants. Figure 3.6 shows that the GA-treated plants are extremely elongated and carry essentially male flowers, which is in contrast to the control seedlings that developed either into male or female plants. The

FIGURE 3.6. Gibberellin-specific induction of maleness in hemp plants; introduction of gibberellin through the roots of the plants. From left to right: 3 females, 3 males (untreated), 5 males (treated).

FIGURE 3.7. 6-BAP-specific induction of femaleness in hemp plants. From left to right: 3 females (untreated), 3 treated plants with hermaphroditic flowers.

treatments with 6-BAP or IAA prevented the formation of any purely male plants. Instead, a large number of plants turned out to be intersexes, i.e., plants carrying flowers that displayed both female and male characteristics (Figure 3.7). ABA treatment had a similar but less marked effect, since it led to a noticeable decrease in the number of male plants and to the formation of a large number of intersexes.

Thus the introduction of growth regulators through the root system at an early developmental stage leads to considerable shifts in the ratio of male to female hemp plants. Gibberellin favors the development of male plants, while 6-BAP and indol-3-acetic acid lead to an increase in the number of intersexes and female plants.

Similar experiments were conducted on spinach (*cv.* Victoria). The seeds were germinated on wet filter paper for 5 days, at 20°C, in the dark. The seedlings that had roots of comparable size were then transferred to 1-liter vessels of water and allowed to grow for 2 days. The hormone treatments were the same as in the experiment with hemp (100 spinach plantlets were treated with each hormone). Twenty-four hours after the treatment the plantlets were transferred to boxes containing soil and allowed to grow under long-day conditions (18 hours of artificial light; 0.009 W/cm^2 xenon lamps) until sex expression became apparent.

The experiments showed that GA stimulates the elongation of male plants; 6-BAP significantly inhibits the growth in height of both male and female plants, whereas ABA has a similar but more moderate effect. Plants treated with IAA had a growth rate comparable to that of the untreated controls (Table 3.3). The 6-BAP and ABA delayed bud development by 6 and 3 days, respectively, compared with the controls. Plants treated with GA started bud development 3 days earlier than the controls.

TABLE 3.3. Effect of phytohormones on growth and development of spinach (introduction through the roots).

Treatment	Height of plants (cm)[1]		Beginning of flower bud development
	Males	Females	
Control	14.0	10.0	11/26
GA	25.5	13.0	11/23
6-BAP	6.0	6.0	12/2
IAA	16.5	12.0	11/26
ABA	11.5	6.5	11/29

[1]Average of 10 measurements.

The introduction of phytohormones into the root system of these seedlings had still more clear-cut effects on sex expression (Table 3.4). GA treatments led, as with hemp, to a significant increase in the number of male plants. In contrast, female individuals constituted the majority of plants in the sets treated with 6-BAP, IAA, and ABA. Figure 3.8 shows a typical distribution of male and female plants in the three sets: 1) control: 2 males and 2 females, 2) GA treatment: 3 males and 1 female, 3) 6-BAP: 3 females and 1 male. The growth inhibition due to 6-BAP can be seen in the far right plant in Figure 3.8: the stem that carries the male flower cluster is very short compared with the stems of other male plants.

The results with spinach confirm the evidence obtained previously with hemp. Comparison of the responses of the two species shows that GA enhances male sex expression to a greater extent in hemp, whereas 6-BAP enhances female sex expression to a greater extent in spinach. In any event, it is clear that drastic effects on sex formation can be produced if physiologically active substances are introduced into the root systems of plants at a sufficiently early developmental stage.

Other experiments were conducted to show the effects of introducing phytohormones into the root systems of monoecious plants that carry unisexual flowers (cucumber and corn). The following plant material was used: 1) cucumber plants (*Cucumis sativa* L.), *cv.* Nerosimye; 2) corn plants (*Zea mays* L.), *cv.* Voronezhskaya-76, Odesskaya-10, and VIR-42, a double interstrain hybrid.

TABLE 3.4. Effects of phytohormones on sex expression in spinach (introduction through the roots).

Treatment	Distribution of sexes (%)		
	Males	Females	Intersexes
Untreated	48.3	51.7	0
GA	78.8	16.3	4.9
6-BAP	11.2	86.7	2.1
IAA	20.8	76.0	3.2
ABA	29.0	71.0	0

FIGURE 3.8. Effect of phytohormones on the induction of maleness and femaleness in spinach plants. 1–4 = Controls (1–2 = females; 3–4 = males); 5–8 = GA treatment (5–7 = males; 8 = female); 9–12 = 6-BAP treatment (9–11 = females; 12 = male).

In cucumber, the location and the proportion of male and female flowers on a given plant are not random. Usual strains of cucumber carry a significantly larger number of staminate flowers than pistillate ones (at least early in the season). Moreover, male sex expression is more pronounced on the central shoot than on the side shoots. In corn there are, as a rule, from two to six female flower clusters (ears) for every male flower cluster (tassel).

Experiments on cucumber plants of the Nerosimye cultivar were conducted as follows. Seeds were germinated on wet filter paper at 25°C for 3 days. Seedlings of comparable size were transferred to 1-liter vessels for 2 days, and then treated with the different phytohormones for 24 hours. The experimental design was the same as in the previous experiments. Each phytohormone was introduced into a set of 14 plants. Twenty-four hours after treatment, the plantlets were transferred to soil and allowed to grow in 18-hour days until sex expression became apparent (using artificial lighting at the greenhouse of the Academy of Sciences of the USSR).

As shown in Table 3.5, the growth in height of the cucumber plants is stimulated by both GA and IAA. The introduction of 6-BAP, on the other hand, leads to the epinasty of leaves at initial developmental stages, and to inhibited growth

TABLE 3.5. Effects of phytohormones on the development of cucumber plants.

	Height of plants (cm)[1]				Day of beginning of flowering
Treatment	10/17	11/10	11/15	12/21	(November)
Untreated	5.5	11.0	13.0	36.0	10th
GA	9.0	22.0	26.0	52.0	6th
6-BAP	4.5	8.0	8.5	14.0	15th
IAA	6.5	18.0	20.0	43.0	10th
ABA	5.5	10.0	12.0	30.0	10th

[1]Average of 10 measurements.

FIGURE 3.9. Cucumber plant treated with 6-BAP and carrying pistillate flowers.

throughout the life cycle. Treatment with GA also results in the early formation of tendrils and of the first five leaves. The plants treated with GA started to flower 4 days earlier than the controls; 6-BAP retarded flowering by 5 days (Table 3.5). In all the plants, the first flowers appeared in the axil of the second leaf–they were all male. The first female flower in the plants treated with ABA and IAA (as well as in the control) appeared on November 18, in the plants treated with 6-BAP on November 15, and in the plants treated with GA on November 21.

Control plants gave rise to both male and female flowers, those treated with GA gave rise almost exclusively to male flowers, while the plants treated with 6-BAP gave rise to an unusually large number of female flowers (Figure 3.9).

The total numbers of male and female flowers formed in the axil of each leaf are shown in Table 3.6. The ratios of male to female flowers were 1) control: 4:1, 2) 6-BAP: approximately 1:1, 3) GA: almost exclusively male flowers. The sex ratios in the case of plants treated with IAA and ABA were comparable to those of the controls.

TABLE 3.6. Effects of phytohormones on the formation of male and female flowers in cucumber (12/28/77).

	No. of flowers in the axil of the leaves[1] (axil no.)																	Total no. of flowers	
Treatment	1	2	3	4	5	6	7	8	9	10	11	12	13	14	15	16	17	Number	%
Control	4/-[2]	10/-	6/-	12/2	6/4	8/2	8/4	6/2	6/2	6/2	6/2	4/2						82/22	78.8/21.2
GA	4/-	6/-	8/-	12/-	12/-	4/2	8/-	8/-	8/-	4/-	8/-	8/-	6/-	6/-	-/1	4/-	12/-	118/3	97.5/2.5
IAA	2/2	4/4	12/2	8/2	6/2	4/2	6/2	8/2	5/1	5/0	2/4	-/1	2/2	-/1				64/27	70.3/29.7
6-BAP	1/1	5/8	4/4	3/3	6/5	6/2	6/4	-/2	3/4									34/33	50.7/49.7
ABA	1/-	4/2	6/4	6/2	6/1	6/1	7/1	4/1	6/1	5/1	5/1	3/-	6/2					65/17	79.2/20.8

[1]Average of 10 measurements.
[2]Male/female.

FIGURE 3.10. Effect of phytohormones on the growth of corn (introduction through the roots). 1 = Control; 2 = GA; 3 = 6-BAP.

These experiments with plants in 3 widely separated families, namely hemp, spinach, and corn, indicate that the introduction of GA and 6-BAP into dioecious plants or monoecious plants bearing unisexual flowers leads to the same changes: GA enhances male sex expression and 6-BAP enhances female sex expression.

Similar experiments were conducted on corn (*cv.* Voronezhskya-76 and Odesskaya-10). The seeds were germinated on wet filter paper for 3 days, at 25°C in the dark. Seedlings with roots of comparable size were transferred to a solution of a given phytohormone for a period of 28 hours (the control plants were transferred to water). The experimental design was the same as with hemp, spinach, and cucumber, except that the GA concentration was raised from 25 to 50 mg/L. After phytohormone treatment, the seedlings were planted in boxes of soil. Each phytohormone was applied to a set of 12 plants that were then allowed to grow

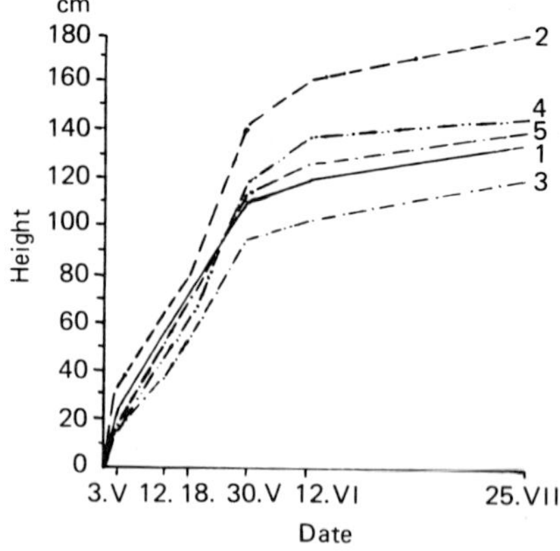

FIGURE 3.11. The effects of phytohormones and inhibitors on growth of corn (introduction through the roots). 1 = Control; 2 = GA; 3 = 6-BAP; 4 = IAA; 5 = ABA.

TABLE 3.7. Effect of phytohormones on the development of corn (introduction through the roots).

Treatment	Beginning of tasseling (June)	Length of tassel (cm)	Ejection of filaments		Length of ear		No. of ears on last plant
			1st ear (June)	2nd ear (July)	1st ear	2nd ear	
Control	10th	24.1	23rd		22.0		1
GA	6th	32.4	28th		13.7		1
6-BAP	15th	15.0	23rd	5th	12.4	19.5	2
IAA	15th	22.2	23rd	5th	17.3	20.6	2
ABA	10th	24.0	23rd		20.8		1

in 18-hour days until sex expression became apparent (lighting: 0.009 W/cm^2 xenon lamps).

It can be seen in Figures 3.10 and 3.11 that GA stimulated the elongation of corn at all stages of development, while 6-BAP had an inhibitory effect. Treatment with IAA and ABA inhibited the growth at early stages, although at harvest the treated plants were slightly higher than the controls. All the plants grown developed 12 internodes. The growth regulators significantly affected the development of the plants. Tasseling in GA-treated plants began 4 days earlier than in the controls, while the appearance of silks was retarded by 5 days (Table 3.7). The tassels of corn plants treated with GA were 8 cm longer and more heavily branched than those of controls. The introduction of 6-BAP and IAA retarded the formation of the tassels by 5 days but did not change the time of appearance of silks.

FIGURE 3.12. Ears formed within the tassels of plants treated with a solution of 6-BAP.

Although the effects of phytohormone treatments in corn are not as obvious as with hemp or spinach, still GA, 6-BAP, and IAA lead to the type of changes that could be expected based on responses of the other species. Specifically, GA stimulates the formation of tassels, while 6-BAP and IAA stimulate the formation of ears. The effect of 6-BAP was particularly well demonstrated in corn of *cv.* Odesskaya-10. In this case the phytohormone (concentration 15 mg/L) was first introduced into the seedlings through the root system, and the tips of the plants were treated again at the 3-leaf stage. The two successive treatments resulted in a dramatic enhancement of female sex expression: 28 of 36 plants developed ears within the tassels (Figure 3.12).

Since the experiments show that the introduction of 6-BAP or IAA into corn plants, either through the roots or by spraying the leaves, leads to the very unusual formation of ears in the tassels, the natural location of ears on the lower part of the stem is perhaps related to the fact that cytokinins are synthesized in the roots and are accumulated in the lower part of the stem, where they participate in the process of female sex expression. Conversely, gibberellins are probably synthesized in the leaves and from thence transported to stem apices where they induce male sex expression.

Overall, the experiments demonstrate that phytohormones introduced into young seedlings can lead to substantial changes in growth rate, time of flowering, and sex ratio, both in dioecious plants and in monoecious plants bearing unisexual flowers. The GA induces male sex expression, while 6-BAP, and to a lesser extent IAA, induce female sex expression.

4
The Roles of Individual Organs and of the Phytohormones They Synthesize in Controlling Sex Expression in Plants

A. An Integral Model of Sex Expression in Plants

To elucidate the roles that different organs (and phytohormones synthesized therein) may play in sex expression, a specific experimental strategy needed to be developed. The first step consisted in removing the roots of young seedlings at the root collar level—dioecious plants and monoecious plants carrying unisexual flowers were used. This first operation was necessarily performed before any sexual differentiation in the stem apices. The seedlings were then transferred to Knop's solution. To locate the organs in which the processes determining the type of sex expression occurred, the seedlings were divided into three groups: 1) with leaves and adventitious roots regenerated *de novo*, 2) with leaves, but without adventitious roots (systematic ablation), and 3) without leaves but with the adventitious roots, all leaves being removed with the exception of the 2–3 youngest apical leaves. The seedlings of all three groups were subjected to a short-term treatment with phytohormones and/or inhibitors by immersing the lower ends of the stems in a solution of the test substance. These treatments were expected to show what internal changes actually control sex expression. The complete experimental design serves as an integral model of sex expression in plants (Figure 4.1).

B. The Roles of Roots, Leaves, and Their Phytohormones in Sexual Differentiation in Dioecious Plants

The role that the different organs may play in plant sex formation has been approached indirectly by several researchers for a long time. Although no specific experiments have been conducted, a number of hypotheses have been put forward. Based on the experiments on cucumber leaves in Chapter 3, in which CO was shown to increase the number of female flowers (Minina and Tylkina 1947), Minina (1952) proposed that the leaves play a determining role in sex expression. On the other hand, Molotkovskyi (1956, 1957, 1960, 1965, 1966,

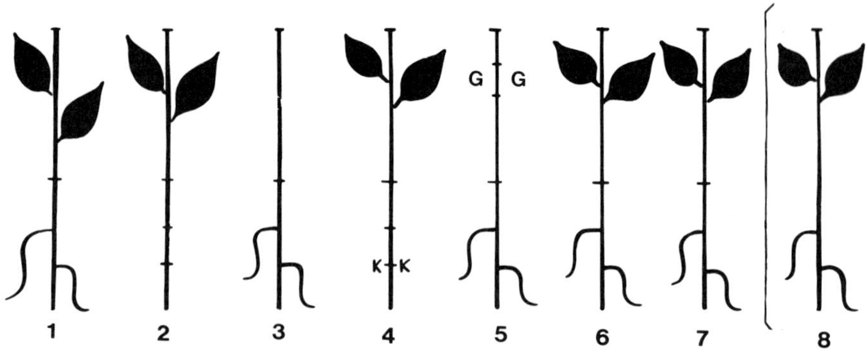

FIGURE 4.1. Integral model of sex expression in plants. 1 = Leaves plus roots; 2 = leaves minus roots; 3 = roots minus leaves; 4 = leaves minus roots plus 6-BAP; 5 = roots minus leaves plus GA; 6 = leaves plus roots plus 6-BAP; 7 = leaves plus roots plus GA; 8 = leaves plus roots (the plant remaining in soil). G = GA, K = 6-BAP.

1968, 1974, 1975),[1] who carried out a number of experiments on the polarity of corn plants as well as on the formation of stems, leaves, and roots, came to the conclusion that the shoot is the organ that essentially determines the formation of the androecium, i.e., of the stamens and anthers, while the root system determines the formation of the gynoecium, i.e., the stigma, style, and ovary. A table showing the comparative measurements of the root system of male and female hemp plants is presented in Table 4.1 (from Dobrunov 1935).

The table shows that the root system of mature female hemp plants is almost three times the size of the root system of the male plants (Dobrunov, 1935). According to Molotkovskyi (1976), at the flowering stage, the "polarity coefficient" of hemp, i.e., the ratio of the shoot to the root, is 5.19 in male plants, and 2.61 in female plants.[2]

It thus appeared logical to the authors to try to determine the specific role of each of the organs in plant sex expression. The initial experimental material was hemp (*Cannabis sativa*).

A preliminary experiment was conducted in the Botanical Gardens of the Penza Pedagogical Institute. Hemp plants (*cv.* US-6) were pregrown in soil until they reached the stage of 2–3 leaf-pairs. In two subsequent experiments, plants were

[1]Molotkovskyi and Sokol (Fiziol. Rast. 25:1290–1292, 1978) also give data on the differences in proteins and sugars among male and female plants of both Gingko and Asparagus [Ed.].

[2]See Lloyd D.G., Webb C.A. (Bot. Revs. 43:177–216, 1977) for a general review of secondary sex characteristics in plants; also Halevy A.H. (Light and the Flowering Process. In: Brue D.V., Thomas B., Cockshull K.E., eds.: Academic Press, 1984) for several instances of the relation between sexuality and the biomass of roots, corms, and bulbs [Ed.].

TABLE 4.1. Root system of hemp plants.

	Male plants	Female plants
Volume of roots (cm^3)	2.90	8.25
Weight of roots (g)	0.25	0.66
Adsorptive surface (m^2)	5.05	13.90

pregrown in soil in 18-hour days in the greenhouse of the Timiryazev Plant Physiology Institute of the Academy of Sciences of the USSR.

The first stage of the experiments consisted of cutting the seedlings at the level of the root collar. The seedlings were first transferred to vessels containing water, then, after 1 day, to a tenfold diluted Knop's solution, after 2 days, to half-strength Knop's solution, and finally, after 2 additional days, to full-strength Knop's solution that was aerated daily. The vessels containing the plants of the first experiment were located in the greenhouse of the Botanical Gardens of the Penza Pedagogical Institute, while the plants of the two other experiments were placed in the growth chambers of the Timiryazev Institute. The conditions were: 16 hours of artificial lighting with 0.002 W/cm^2, temperature: 20°C, relative humidity: 80% (Figure 4.2). The plants were later divided into two sets: 1) with leaves and adventitious roots (leaves plus roots), and 2) with leaves, but adventitious

FIGURE 4.2. General view of the experiment. 1 and 2 = Experimental plants with roots removed; 3 and 4 = control plants with roots present.

TABLE 4.2. Sex expression in hemp plants grown in natural daylight (ND) and long days (LD).

Planted in	No. of experiments	Photoperiod	Total no. of plants	Male plants		Female plants	
				No.	%	No.	%
Bed (25 m²)	1	ND	2970	1390	46.8	1580	53.2
Boxes	1	LD	91	48	52.7	43	47.3
Boxes	2	LD	80	43	53.7	37	46.3

roots systematically removed (leaves minus roots). At the beginning of each experiment, the apices of the plants were carefully examined to be sure that they were still at the vegetative stage. The number of male and female plants remaining in the beds (experiment 1) or in the boxes in LD (experiments 2 and 3) were similar (Table 4.2).

The first (preliminary) experiment yielded the following results: of 40 plants initially in each group, only 35 with leaves and roots and 26 with leaves but no roots remained at the end of the experiment. The systematic removal of roots led to a significant increase (76.9%) in the number of male plants, while the presence of roots enhanced female sex expression (80.0% increase) (Table 4.3). The plants started flowering at the same time in both sets, and the female plants eventually set seeds.

In the second experiment, with 70 plants in each set, 64 survived in the set with leaves and roots, and 60 in the set with leaves but no roots. In the third experiment, of 100 plants in each set, 91 survived with leaves and roots and 90 with leaves but no roots. The results obtained in these two last experiments with larger numbers of plants are entirely consistent with the results of the first experiment (Table 4.3 and Figure 4.3).

Thus it is clear that both roots and leaves play an important role in sex formation in hemp plants: the roots promote female sex expression, and the leaves promote male sex expression. The effect of root removal on apical development is

TABLE 4.3. Influence of roots and leaves on sex expression in hemp (leaves present on all plants).

Treatment	No. of plants	Male plants		Female plants	
		No.	%	No.	%
Intact	35	7	20.0	28	80.0
Roots removed	26	20	76.9	6	23.1
Intact	64	5	7.8	59	92.2
Roots removed	60	53	88.3	7	11.7
Intact	91	8	8.8	83	91.2
Roots removed	90	81	90.0	9	10.0

FIGURE 4.3. Roles of roots and leaves in sex expression in hemp. 1 = Plants bearing leaves, but with roots removed; 2 = plants with developed adventitious roots.

not irreversible: if the removal of adventitious roots was stopped at the beginning of the differentiation of male flower primordia, mixed flowers carrying male and female characteristics appeared in the flower clusters (Figure 4.4).

To establish whether the results obtained with hemp (a SD dioecious species) are also characteristic for LD dioecious species, similar experiments were carried out with spinach. Spinach plants were initially grown in soil in 18-hour days (in the greenhouse of the Plant Physiology Institute of the Academy of Sciences of the USSR). When the seedlings reached the stage of the third visible leaf, some were cut off at the level of the root collar and divided into two sets (leaves plus (adventitious) roots, and leaves but no roots); they were then transferred to Knop's solution. The remaining seedlings developed into 52 adult male plants (47.3% of the total) and 58 adult female plants (52.7% of the total).

Table 4.4 and Figure 4.5 show that the regeneration of roots results in a considerable increase in the number of female plants, while their systematic removal

FIGURE 4.4. Role of roots and leaves in sex expression in hemp plants. 1 = With adventitious roots (female plant); 2 = after derooting, but adventitious roots left—the beginning of male sex expression (intersex plant); 3 = all roots systematically removed (male plant).

leads to enhanced male sex expression. The regeneration of adventitious roots in spinach could be detected as soon as 3 to 6 days after their removal, and bud development began in all plants 7 to 10 days after the formation of the new roots.

Thus the results with spinach confirm those obtained with hemp, namely that the formation and activity of the roots are responsible for female sex expression, while the leaves play a major role in male sex expression.

TABLE 4.4. Influence of roots and leaves on sex expression in spinach (leaves present on all plants).

Treatment	Total no. of plants	Male plants		Female plants	
		No.	%	No.	%
Intact	103	7	15.0	96	85.0
Roots removed	100	85	85.0	15	15.0

Further experiments were designed to link the roles of roots and leaves in sex expression to specific physiologically active compounds. In so doing, arguments put forward by other researchers, as well as some of the original experiments presented in Chapter 3, were taken into consideration. Specifically, based on the fact that the introduction of 6-BAP through the roots of young seedlings of hemp and spinach leads to enhanced female sex expression, it was argued that the introduction of cytokinins into plants without roots would have a similar effect, i.e., induce femaleness. This hypothesis was supported by studies on the synthesis in roots of hormonally active substances belonging to the cytokinin group (Sabinin 1949; Mothes et al. 1959; Kulaeva 1962; Mothes 1964).

The experiment aimed at testing this idea was similar to those on the roles of leaves and roots in sex expression. The plants were cut and transferred to vessels

1 2 3

FIGURE 4.5. Role of roots in sex expression in spinach. 1 = Roots removed, addition of 6-BAP; 2 = roots present; 3 = roots removed (controls for No. 1).

TABLE 4.5. Influence of cytokinin on sex expression in hemp and spinach plants deprived of roots (leaves present on all plants).

Treatment	Total no. of plants	Male plants No.	Male plants %	Female plants No.	Female plants %
Hemp					
Roots removed	44	34	77.2	10	22.8
Roots removed, 6-BAP added	48	9	18.8	39	81.2
Spinach					
Roots removed	100	85	85.0	15	15.0
Roots removed, 6-BAP added	98	16	16.4	82	83.6

containing either water or a solution of 6-BAP (15 mg/L). After 28 hours, all the plants were transferred to Knop's solution. The experiment included two sets of plants: 1) leaves minus roots, 2) leaves minus roots + 6-BAP. Each set included, with hemp, 50 plants, and with spinach, 100 plants.

The results are presented in Table 4.5 and Figures 4.6 and 4.7. The introduction of 6-BAP into freshly derooted hemp or spinach plants leads to a significant

1 2

FIGURE 4.6. Influence of cytokinin on sex expression in hemp plants with roots removed. 1 = Control plants (males); 2 = experimental female plants treated with 6-BAP.

FIGURE 4.7. Role of roots and 6-BAP in sex expression in spinach seedlings. 1–3 = Females, roots removed, treated with 6-BAP; 4–6 = females with roots; 7–9 = males, roots removed.

increase in the number of female individuals. Thus, the synthesis of cytokinins in the roots is definitely relevant to the role that roots play in female sex expression.

In view of this result, it was expected that the introduction of GA into plants that lacked leaves would lead to an enhanced formation of male flowers. Such a result would be consistent with the experiments on young seedlings in Chapter 3, and more generally with the studies on the synthesis of GA in leaves (Chailakhyan 1958, 1968, 1971, 1972; Stoddart and Lang 1967).

The experiment on the influence of GA was carried out in the same manner as above. There were three sets of plants: 1) roots plus leaves, 2) roots minus leaves, and 3) roots minus leaves plus GA. The seedlings were treated with a solution of GA_3 (25 mg/L) for 28 hours. In sets 2 and 3 the leaves were removed only after the regeneration of adventitious roots, which required 10 to 12 days for hemp, and 3 to 6 days for spinach. In hemp, bud development was initiated 17 to 19 days after the beginning of the experiment; in spinach, the corresponding time was 9 to 11 days. The development of adventitious roots in sets 2 and 3 was delayed due to the systematic removal of leaves.

TABLE 4.6. Influence of GA on sex expression in hemp and spinach plants with and without leaves (roots present on all plants).

	Hemp				Spinach			
	Male plants		Female plants		Male plants		Female plants	
Treatment	No.	%	No.	%	No.	%	No.	%
Leaves present	15	17.6	70	82.4	16	15.3	89	84.7
Leaves removed	18	20.0	72	80.0	14	13.9	87	86.1
Leaves removed, GA added	68	80.9	16	19.1	76	76.7	23	23.3

FIGURE 4.8. Role of leaves in sex expression in hemp. 1. Leaves and roots present (the majority are females); 2. roots only (the majority are females); 3. roots only, GA added (the majority are males).

The results are presented in Table 4.6 and Figures 4.8 and 4.9. As observed earlier, the sex ratio in plants that retained their leaves and formed adventitious roots (i.e., leaves plus roots) was shifted toward the female sex, 82.4% (hemp) and 84.7% (spinach) of the plants being female. A comparable shift was observed

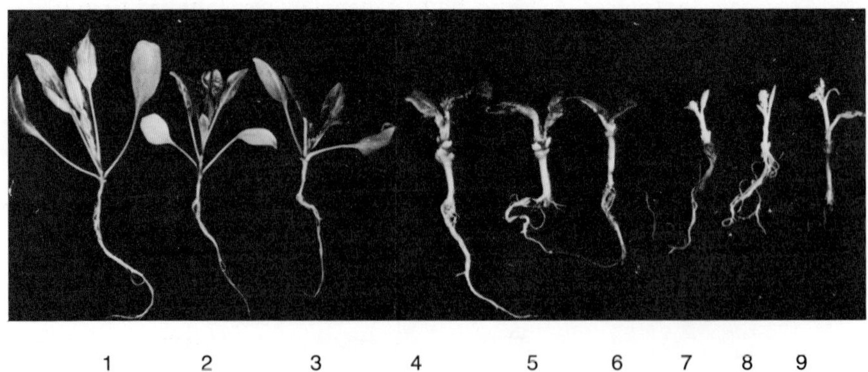

FIGURE 4.9. Role of leaves in sex expression in spinach. Roots present on all plants. 1–3. Control plants with leaves intact (1–2 are females, 3 is male); 4–6. leaves removed (all females); 7–9. leaves removed, GA added (all males).

in those plants that lacked their leaves only (i.e., roots minus leaves), the proportion of female plants being 80.0% in hemp, and 86.1% in spinach. In contrast, the majority of plants deprived of their leaves and treated with GA (roots minus leaves + GA) developed into male individuals (80.9% and 76.7%). Thus, as long as the roots are present, the number of female plants is increased, regardless of whether the leaves are removed or not (sets 1 and 2). However, if at the time of the first removal of leaves, the plants were treated with GA, *male* sex expression was favored (set 3). This result demonstrates that it is indeed the gibberellins synthesized in the leaves of hemp and spinach that play a determining role in the formation of male sex organs.

In an essentially identical experiment, the effect of GA was enhanced by removing the leaves of experimental hemp seedlings at the time of GA treatment, i.e., before any regeneration of roots had occurred. Such an early removal of leaves in sets 2 and 3 (roots minus leaves and roots minus leaves plus GA) significantly delays the regeneration of roots, for it was 24 days after the beginning of the experiment when new roots could first be observed. Also their growth in length was extremely poor. Bud development in GA-treated plants (set 3) was very rapid (9 days after GA treatment) compared to that in the other plants (17 and 21 days after the beginning of the experiment in sets 1 and 2, respectively). The detailed results are presented in Table 4.7. The overwhelming majority (89.5%) of seedlings treated with GA (set 3) developed into male plants, while most of those that retained their leaves and adventitious roots (set 1) developed, as expected, into female individuals (86.2%). The early removal of leaves in set 2 (roots minus leaves) led to the formation of an equal number of male and female plants (52.6% males and 47.4% females). This result can be explained by the delayed appearance of adventitious roots; because budding occurs before root regeneration, the female sex expression promoted by the roots is not apparent.

The experiments presented above demonstrate that the synthesis of gibberellins in the leaves of dioecious plants is an important factor in the role that leaves play in male sex expression.

Further experiments were directed at determining the roles of the same phytohormones—gibberellins and cytokinins—in sex expression of intact dioecious plants (hemp and spinach), i.e., plants that retained their leaves and adventitious roots.

TABLE 4.7. Influence of GA on sex expression in hemp plants with leaves removed (roots present on all plants).

Treatment	Total no. of plants	Male plants		Female plants	
		No.	%	No.	%
Leaves present	102	14	13.8	88	86.2
Leaves removed	95	50	52.6	45	47.4
Leaves removed, GA added	105	94	89.5	11	10.5

TABLE 4.8. Influence of cytokinin and gibberellin on sex expression in hemp and spinach plants (all plants intact).

	Hemp				Spinach			
	Male plants		Female plants		Male plants		Female plants	
Treatment	No.	%	No.	%	No.	%	No.	%
None, controls	15	17.6	70	82.4	16	15.3	89	84.7
6-BAP	12	14.2	73	85.8	9	9.0	92	91.0
GA	75	89.3	9	10.7	90	87.3	13	12.7

Spinach and hemp plants were pregrown as described. The freshly cut seedlings were transferred initially into water, and then into Knop's nutrient solution. The experiment was conducted in a growth chamber (16-hour days, temperature 20°C, humidity 80%). There were three sets of plants: 1) leaves plus roots, 2) leaves plus roots plus 6-BAP (the seedlings were treated with a 15-mg/L solution for 24 hours), and 3) leaves plus roots plus GA (25 mg/L for 24 hours).

Regeneration of adventitious roots was observed after 10 to 12 days in the hemp plants, and after 3 to 6 days in the spinach. The relevant plants (sets 2 and 3) were then treated with the appropriate phytohormones. The time from the

FIGURE 4.10. Effects of 6-BAP and GA on sex expression in hemp. 1 = Control plants (majority are females); 2 = 6-BAP (the majority are females); 3 = GA (the majority are males).

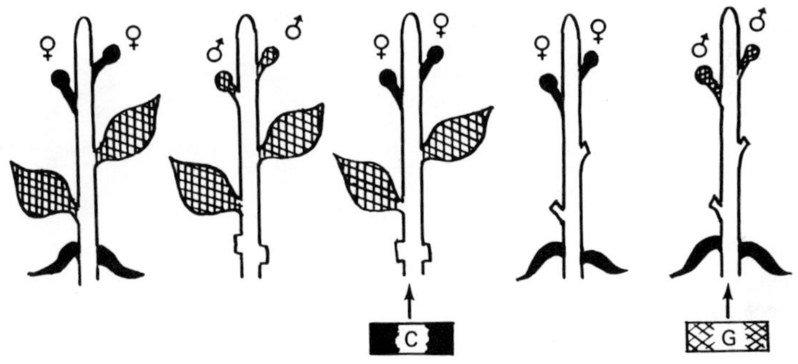

FIGURE 4.11. Roles of organs and phytohormones in sex expression in plants. C = cytokinins; G = gibberellins.

beginning of the experiment to bud development varied as follows: hemp set 1 = 21 days, set 2 = 19 days, set 3 = 17 days; and spinach set 1 = 11 days, set 2 = 9 days, and set 3 = 8 days.

The results (Table 4.8 and Figure 4.10) show that both in hemp and in spinach, the untreated seedlings (set 1) developed predominantly into female plants (82.4% in hemp, 84.7% in spinach). The proportion of resulting female individuals was even slightly higher among the plants of set 2 treated with 6-BAP, (85.8% in hemp, 91.0% in spinach). The relatively small effect of 6-BAP may perhaps be explained by the already large amounts of endogenous cytokinins that must have been synthesized in the adventitious roots.

In strong contrast, the treatment with GA (set 3) led to the development of predominantly male plants (89.3% in hemp, 87.3% in spinach).

Based on these experiments and on earlier research, we conclude that sex expression in dioecious plants is an interactive process between the phytohormones and the organs in which they are synthesized (Figure 4.11). The phytohormones tested, 6-BAP and GA, are evidently easily taken up by young plants from weak aqueous solutions. They induce the same changes regardless of the mode of uptake, whether via primary roots, sectioned stems, or regenerated adventitious roots. In all cases, 6-BAP causes an increase in the number of female plants, while GA_3 favors the formation of male individuals.

Normally, both in hemp and spinach, sex expression in intact plants is regulated so as to produce overall about the same number of male and female individuals. The removal of primary roots, followed by root regeneration and leaf development, causes enhanced female sex expression. However, if the systematic removal of roots is coupled with the introduction of a cytokinin (6-BAP), this results in a further considerable increase in the number of female plants. This is true both for hemp and spinach. Finally, the removal of leaves only, if coupled with the introduction of GA, causes a large increase in the number of male individuals. Thus, sex expression in two dioecious plants, hemp and spinach,

TABLE 4.9. Influence of leaves, roots and phytohormones on growth and development of cucumber plants.

Treatment	Height of plants (cm)[1]			Beginning of flowering
	November 10	November 15	December 24	
Intact	14.0	22.0	75.0	November 10
Roots removed	4.0	4.0	8.5	November 11
Leaves removed	3.0	3.0	6.0	November 26
Roots removed, 6-BAP added	3.5	3.5	7.5	November 12
Leaves removed, GA added	4.0	4.0	11.0	November 21
Intact, 6-BAP added	6.0	12.0	30.0	November 14
Intact, GA added	18.0	29.0	95.0	November 7
Plants left in soil	11.0	13.0	36.0	November 10

[1]Average of 5 measurements.

involves leaves that behave like organs synthesizing GA, and roots that behave as organs synthesizing cytokinins.[1]

C. The Roles of Individual Organs and Their Phytohormones in Sex Expression in Monoecious Plants Carrying Unisexual Flowers

The findings regarding the action of roots and leaves in controlling sex expression in two dioecious plants, namely hemp and spinach, led to further experiments on monoecious species that carry unisexual flowers, i.e., cucumber plants (*cv.* Nyerosimye). Experimental plants were pregrown[2] in 18-hour days in boxes of soil. As soon as the first leaf appeared (apices at the vegetative stage), most of the seedlings were cut at the level of the root collar and transferred to vessels initially containing water, and later, Knop's solution. The seedlings were grown in growth chambers in 16-hour days, at 20°C, and in 80% relative humidity. The seedlings were divided into 7 sets of 5 to 7 plants each. An additional group of 5 to 7 plants remained in the boxes; their growth and development were also monitored (Table 4.9).

[1]Such sex-determining influences are not limited to leaves and roots. Trivers (Social Evolution, Menlo Park, California, Benjamin/Cummings Publishers Co., 1985, p. 289) states that in *Rumex*, when the pollen supply is sparse or normal the sex ratio of the resulting seedlings is nearly 1:1; however, when there is excess pollen, the female:male ratio is 5:1. The data of this and following chapters suggest that the pollen is rich in cytokinin [Ed.].

[2]Location: the greenhouse of the Timiryazev Plant Physiology Institute of the Academy of Sciences of the USSR.

FIGURE 4.12. Roles of organs in sex expression in cucumber plants. 1 = Leaves plus adventitious roots; 2 = leaves minus roots; 3 = leaves minus roots plus 6-BAP.

The roles of roots and leaves in sex expression were studied as follows. In set 1, the plants were allowed to regenerate their roots (leaves plus roots); in set 2, the roots were systematically removed (leaves minus roots); and in set 3, the new roots were left, but the leaves were removed (roots minus leaves). The effects of hormones were studied in sets 4, 5, 6, and 7. In set 4, 6-BAP was introduced by soaking the stems of freshly harvested seedlings in a 15-mg/L solution for 24 hours; the adventitious roots were systematically removed (leaves minus roots plus 6-BAP). The plants treated with GA_3 (roots minus leaves plus GA_3, set 5) were first allowed to regenerate their roots, then deprived of their leaves, and soaked in a 25-mg/L solution for 24 hours. In set 6 (leaves plus roots plus 6-BAP) and set 7 (leaves plus roots plus GA_3), the hormones were introduced after root regeneration. The general experimental set-up is shown in Figure 4.12. The regeneration of adventitious roots of the cucumber plants began 5 to 6 days after transfer to Knop's nutrient solution.

The results on the growth of the plants are presented in Table 4.9. Those that were allowed to develop both leaves and adventitious roots (set 1) were twice as high as the plants that remained in the soil. The simple addition of GA_3 (set 7) stimulated elongation, while 6-BAP (set 6) had an inhibitory effect. Growth was dramatically reduced in plants that lacked roots, especially in those that lacked leaves (set 3).

The plants growing in soil and those in set 1 initiated flowers at the same time. However, the plants in set 5 (leaves plus roots plus GA_3) flowered 3 days earlier than in set 1. On the other hand, the introduction of 6-BAP (set 4) delayed flowering by 4 days. The absence of leaves alone (set 3) led to very late flowering (16 days after set 1). The introduction of GA_3 (set 5) reduced the delay caused by the absence of leaves by 5 days. In all other cases, no significant changes in flowering were observed.

Table 4.10 summarizes the sex expression in the different sets of plants. The removal of roots only (set 2) results in the formation of *exclusively* male flowers, while the removal of leaves only (set 3) very strongly favors the appearance of female flowers (Figures 4.13A, 4.14A). This finding indicates that it is indeed the leaves and the roots themselves that are responsible for the formation of male and female flowers, respectively. The introduction of 6-BAP into plants that lack their roots (set 4) leads to an increase in the number of female flowers of up to 45.7% (Figure 4.13B). As expected, GA has the opposite effect: 98.6% of flowers in plants lacking their leaves and treated with GA_3 (set 5) turned out to be male. The introduction of the phytohormones after root regeneration (sets 6 and 7) had the expected effects: 6-BAP and GA induce, respectively, female and male sex formation.

Thus, the role that roots play in female sex expression involves the synthesis of cytokinins in the roots; correspondingly, the effect that leaves have on male sex expression is related to the synthesis of gibberellin in the leaves.

The experiments also show that the roles played in sex expression by individual organs and by the phytohormones they synthesize are the same in a monoecious species carrying unisexual flowers (cucumber), as they are in dioecious species (hemp and spinach).

D. Individual and Combined Effects of Hormones and Growth Inhibitors on Sex Expression in Hemp

Sex expression in plants is not only affected by cytokinins and gibberellins. As indicated in Chapter 3, a number of other phytohormones, e.g., auxins, ethylene, abscisic acid, and some growth inhibitors such as the CCC retardant, are known to play important roles (Frankel and Galun 1977; Sidorskyi 1978). It is also known that certain natural and synthetic inhibitors and retardants act as either partial antagonists (Turetskaya et al. 1969) or total antagonists of a given phytohormone (Champault 1973; Muromtsev and Khrianin 1974; Jaiswal and Mohan Ram 1974; Ivanova 1974; Chailakhyan and Nekrasova 1976; Friedlander et al. 1977).

The following experiments compare the effects of inhibitory and stimulatory phytohormones. The integral model of sex expression was the rationale for the experimental design. As in previous experiments, hemp plants (*Cannabis sativa* L., cultivar US-6) were pregrown in soil in 18-hour days (location: the

TABLE 4.10. Sex expression in cucumber plants as a function of treatment and growth.

Treatment	No. of flowers, male/female, in each axil (axil no.)																	Total (male/female)	
	1	2	3	4	5	6	7	8	9	10	11	12	13	14	15	16	17	No.	%
Control, intact	2/-	8/-	10/-	12/2	10/4	10/2	4/2	10/-	10/-	10/-	10/2	10/2	8/2	-/4				114/20	85.0/15.0
Roots removed	3/-	6/-	12/-	12/-	6/-	6/-	6/-	6/-										63/-	100/-
Leaves removed	-/2	-/3	-/5	-/6	-/6	-/8	2/8											2/38	5.0/95.0
Roots removed, 6-BAP added	1/-	3/2	3/3	4/3	2/4	3/2	3/2											19/16	54.3/45.7
Leaves removed, GA added	4/-	6/1	10/-	6/-	8/-	6/-	8/-	6/-	4/-	8/-	6/-							72/1	98.6/1.4
Intact, 6-BAP added	2/4	2/4	4/4	6/4	7/5	-/5	5/4	5/3	6/3	6/3	4/4	3/3	4/6					54/52	50.9/49.1
Intact, GA added	7/-	12/-	11/-	16/1	20/1	10/-	8/1	12/-	12/-	10/-	16/-	10/-	14/-	7/1	12/2	14/-	6/-	197/6	97.0/3.0
Plants remaining in soil	4/-	10/-	6/-	12/2	6/4	8/2	8/4	6/2	6/2	6/2	6/2	4/2						82/22	78.8/21.2

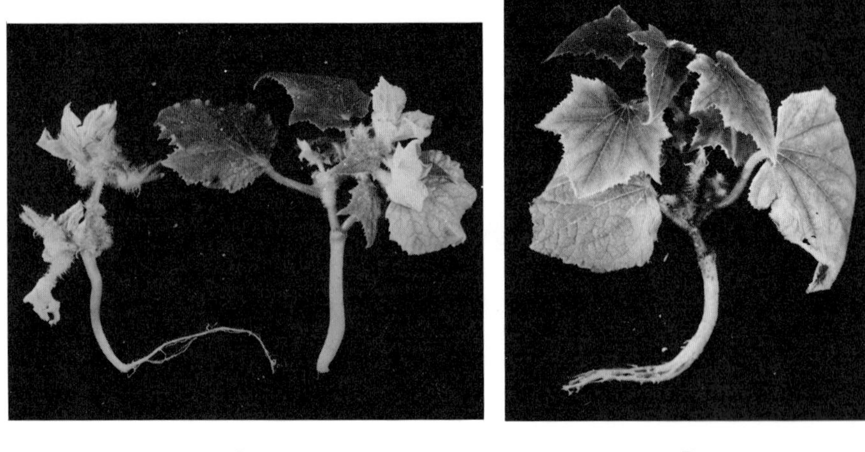

FIGURE 4.13. Roles of individual organs and of a cytokinin in sex expression in cucumber plants. (A) Left: leaves removed, the majority of flowers are pistillate; right: roots removed, the majority of flowers are staminate. (B) Roots removed, 6-BAP added, all the flowers are pistillate.

Greenhouse of the Plant Physiology Institute of the Academy of Sciences of the USSR). At the stage of 2 to 3 pairs of leaves, the plants were cut at the level of the root collar and transferred to vessels containing a solution of a given phytohormone for 24 hours. The plants were then allowed to grow in 16-hour days in artificial light (0.002 W/cm^2, temperature 20°C, relative humidity 80%). In one set of experiments, where the individual and combined effects of GA and CCC were tested, the plants were allowed to regenerate their roots, but the leaves (with the exception of the two apical ones) were systematically removed. In the other set, the leaves remained but the adventitious roots were systematically removed. In this case, the individual effects of 6-BAP, IAA, ABA, CCC, and Ethrel (2-chloroethylphosphonic acid) and the combined effect of 6-BAP and ABA were studied.

The experimental plants were treated in 10 sets as follows: 1) control (water); 2) GA_3, 25 mg/L, no leaves; 3) CCC, 8 mg/L, no leaves; 4) GA_3, 25 mg/L plus CCC, 6 mg/L added 24 hours later, no leaves; 5) 6-BAP, 15 mg/L, no roots; 6) IAA, 15 mg/L, no roots; 7) ABA, 10 mg/L, no roots; 8) CCC, 8 mg/L, no roots; 9) Ethrel, 0.01%, no roots; 10) 6-BAP, 10 mg/L plus ABA, 5 mg/L added 24 hours later. The optimal concentrations of the different substances had been determined in preliminary experiments. Each set included 112 plants. These experiments were repeated 8 times.

Table 4.11 summarizes the results obtained. One of the effects of GA_3 was early bud development (8 days before the controls). It is possible to counteract this effect of GA by treatment with CCC, although CCC delays bud development only

FIGURE 4.14. Roles of individual organs and of a gibberellin in sex expression in cucumber plants. (A) Roots removed, the majority of flowers are staminate. (B) Leaves removed, GA added, the majority of flowers are staminate.

if the roots are present. IAA and Ethrel delayed bud development by 2 and 5 days, respectively. In all other sets, the buds grew out at the same time as in the control plants. Ethrel had a deleterious effect on the leaves: 2 to 3 days after the beginning of the experiment they turned yellow, and by 10 days all but the two apical leaves had fallen.[1] The axillary buds gave rise to new leaves that were thin and revolute and were clearly demonstrating epinasty.

The results confirmed that GA determines male sex expression in leafless hemp plants, and that 6-BAP determines female sex expression in plants that lack roots (Table 4.11). IAA and ABA treatments on plants deprived of roots had the same

[1]This exhibits the well-known effect of ethylene in promoting senescence and leaf abscission in some species [Ed.].

TABLE 4.11. Effects of sex growth regulators on budding and sex expression in hemp.

Treatment	Beginning of flower bud formation	Percentage of plants		
		Males	Females	Intersexes
Control	April 3	19.0	81.0	
Leaves removed, GA added	March 26	85.2	14.8	
Leaves removed, CCC added	April 5	25.8	74.2	
Leaves removed, GA + CCC added	March 30	70.9	29.1	
Roots removed, 6-BAP added	April 3	16.4	83.6	
Roots removed, IAA added	April 5	28.1	66.6	5.3
Roots removed, ABA added	April 3	39.3	57.4	3.3
Roots removed, CCC added	April 3	32.0	68.0	
Roots removed, Ethrel added	April 8		54.1	45.9
Roots removed, 6-BAP + ABA added	April 3	34.4	65.6	

effect on sex expression as when the substances were simply introduced through the roots of seedlings, namely, treatments with IAA and, to a smaller extent, ABA led to a small increase in the number of female plants and to the appearance of intersexes. It should be noted that if ABA simply is sprayed on the plants, it does not affect sex expression in hemp (Mohan Ram and Jaiswal 1972).

Sex expression in the experimental plants was strongly affected by the introduction of 0.01% Ethrel (set 9); a higher concentration of Ethrel, i.e., 0.05%, caused the death of the plants. Set 9 did not include any exclusively male plants. Instead, a number of intersexes had developed (Figure 4.15). In these plants, male and female flowers occupied the lower part of the flower cluster, hermaphroditic flowers developed in the middle, and female flowers appeared in the upper part. On the average, each plant carried 23 female, 6 male, and 6 hermaphroditic flowers. Interestingly, the stigmas of female flowers on all the Ethrel-treated plants were significantly elongated, compared with those in the control plants.

The growth retardant CCC caused an increase in the number of female plants. The effect was slightly stronger in plants that had regenerated their roots (74.2% females) than when the roots were systematically removed (68.0% females). These results may be explained by the inhibitory effect that CCC has on GA synthesis (Carr and Reid 1967; Ivanova 1971) and the accompanying increase in cytokinin content (Skene 1968).

FIGURE 4.15. Flower cluster of a hemp plant (intersex) treated with Ethrel.

Comparison between individual and combined effects of phytohormones and inhibitors (sets 2–5, 7, 10) revealed that CCC partially inhibits the male sex expression caused by GA, and that ABA counteracts the female sex expression caused by 6-BAP (cf. Figure 4.16). Thus, it appears that CCC and ABA act as antagonists of gibberellins and cytokinins, respectively.[1] In the experiments of Mohan Ram and Jaiswal (1972), the immediate spraying of GA-treated hemp plants with ABA inhibited the action of GA in causing male sex expression. The

[1]That CCC inhibits the biosynthesis of GA has long been known (see the review of Lang A. Ann. Rev. Plant Physiol. 21:637–670, 1970) [Ed.].

FIGURE 4.16. Effects of phytohormones and growth inhibitors on sex expression in hemp. 1 = Control, intact; 2 = leaves removed, CCC added (the majority are females); 3 = leaves removed, GA and CCC added (the majority are males); 4 = roots removed, ABA added; 5 = roots removed, 6-BAP and ABA added.

results of the experiment presented here are entirely consistent with the studies of Champault (1973) on the effects that GA, 6-BAP, IAA, CCC, and other substances (except Ethrel) have on the development of internodes in male and female mercury plants (*Mercurialis annua* L.).

All the experiments clearly indicate 1) that gibberellins and cytokinins are the leading hormones in controlling male and female sex expression in hemp, but 2) that gibberellins and cytokinins interact with other phytohormones and inhibitors. Evidently, the processes of differentiation of sexual characteristics in plants are regulated by a multicomponent hormonal system with a defined stoichiometry. The overall effect on sex expression is dependent on the concentrations of the different hormones and on their interactions with other existing factors.

According to Skoog and Miller (1957) and Kulaeva (1973), the set of hormones regulating different physiological processes may always be the same, but the factor that triggers one of many genomically encoded developmental programs is the concentration of the predominant hormone. In the case of sex expression in dioecious plants and in monecious plants carrying unisexual flowers, the predominant hormones are cytokinins and gibberellins. The capacity of these hormones to direct a given type of sexual differentiation is clearly demonstrated by the introduction of exogenous cytokinins (enhanced femaleness) or exogenous gibberellin (enhanced maleness).

Sex expression, like many other physiological processes in plants, apparently is under the control of a regulatory system composed not only of phytohormones but also of natural inhibitors. Consistent with this hypothesis is the finding that female hemp flower clusters are rich in cytokinins, while the male clusters contain high levels of abscisic acid (Engelbrecht 1973). Similarly, female flowers of cucumber contain more cytokinins than male flowers, as content in inhibitors of phenolic nature are the same in both types of flowers. Male flowers of *Mirabilis jalapa* also contain relatively large amounts of gibberellins (Murakami 1975). The total control of sex expression probably also includes interactions between gibberellins and abscisic acid, as well as between cytokinins and phenolic inhibitors.

5 Differential Content of Phytohormone Activity in Male and Female Plants

To gain a more thorough understanding of the roles played by the phytohormones and by the different parts of the plant in sex expression in dioecious species, we set up experiments aimed at characterizing the biological activity of gibberellins, cytokinins, auxins, and growth inhibitors present in different plant organs. The methods used in these experiments are described in each case. Fresh plant material for analytical work was frozen in liquid nitrogen, freeze-dried, and stored in a desiccator at 4°C. Alternatively, plant tissues and organs were fixed in boiling 96% ethanol.

A. Gibberellins

The content of free gibberellin-like substances (GLS) was determined by the method of Lozhnikova et al. (1967, 1973). The control experiment was conducted on dwarf pea plants (*cv.* Pioneer G-1 and Rannyi Konservnyi No. 21). Etiolated pea seedlings were grown by the method of Muromtsev and Rusanova (1966). The samples (extracts from 10 g of crude material) were subjected to downward paper chromatography (Schleicher and Schüll 2043a) using isopropanol:water (5:2) as solvents and 5% sulphuric acid (0.5 M) as a developer.

B. Cytokinins

The extraction of cytokinins was conducted using the basic methods developed by Letham and Williams (1965) and Van Staden (1973) adapted for endogenous cytokinins by Konopskaya (1977). After extraction of 10 g freeze-dried material with 80% ethanol, the extract was kept at 4°C for 24 hours and re-extracted 3 times for 2 hours each, using a rotating separatory funnel (3 × 80 mL ethanol). The ethanol fractions were then pooled and evaporated on a rotor at 30°C. The aqueous fraction that remained in the separatory funnel was treated with ether (3 × 25 mL) at pH 3 (by addition of 0.1 N HCl) and then with water-saturated N-butanol (3 × 25 mL) at pH 7 (by addition of $NaHCO_3$). The ether fraction was evaporated at room temperature, the aqueous fraction at 40 °C, and the butanol fraction at 45 °C. The cytokinins were eluted from the dry residues with 2 to 3 mL 35% ethanol and chromatographed using Whatman No. 3 paper and butanol: ammonia (4:1). Purified kinetin and zeatin were used as markers. The activity of cytokinins was determined in a bioassay with amaranth seedlings (*Amaranthus caudatus* L.), as described by Kefeli et al. (1977). This method was first introduced by Bigot (1968) and subsequently rearranged and modified by Biddington and Thomas (1973) and Mazin et al. (1976). The concentration of betacyanins was determined by measuring the optical density at 540 nm.

C. Auxins and Growth Inhibitors

The biological activity of auxins and growth inhibitors was determined by the method of Kefeli et al. (1966, 1973). The fixed plant material was extracted 6 times with peroxide-free ether. The ether was evaporated under a stream of cold air, and the residue was redissolved in 2.5 mL of 96% ethanol. The extract (equivalent to 10 g of raw material) was spotted onto chromatographic paper (16 × 36 cm) (Whatman 2- and 3-mm) prewashed in 20% formic acid. Chlorophylls, fats, and resins were removed by washing in running toluol (20 to 40 minutes). The spotted samples were then subjected to upward chromatography with N-butanol:glacial acetic acid:water (40:12:28). One of the chromatographs was examined under ultraviolet (UV) light in ammonia vapor and treated with diazotized sulfanilic acid. The position of the spots was recorded. The other three chromatographs were divided into 10 strips and bioassayed, using wheat coleoptiles (*cv.* Ulyanovka) according to the method of Boyarkin (1966).

The resulting histograms show the physiological activity of gibberellins, cytokinins, auxins, and natural growth inhibitors. The R_f values are plotted on the horizontal axis. The vertical axis includes the growth of the test object, i.e., the length of experimental plants expressed in the percentage of the length of control plants treated with purified gibberellins, auxins, and growth inhibitors, as well as the optical density of the betacyanins (as the percentage of that in the cytokinin control).

D. Results

Initial experiments were aimed at determining the biological activity of the phytohormones and inhibitors as a function of the growth and development of the hemp plants. The results indicate that although the contents and the activities of growth regulators vary with time, they are clearly different in male and female plants. The peak auxin and gibberellin-like activities are found in the leaves of hemp plants at the beginning stage of the segregation of the sexes (Figure 5.1). Suggestively, it is at this very stage that hemp plants experience a dramatic increase in growth, with both the appearance of new internodes and the simul-

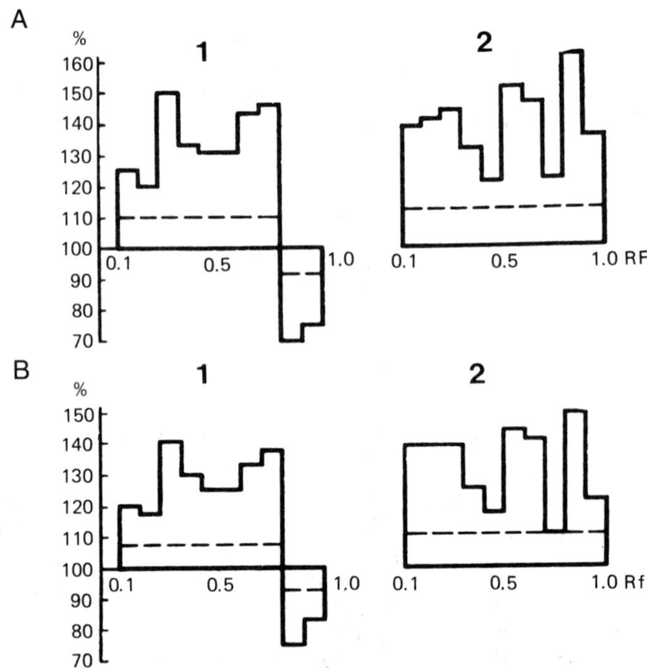

FIGURE 5.1. Biological activities of phytohormones and inhibitors in the leaves of male (A) and female (B) hemp plants at the beginning stage of segregation of the sexes. 1 = Auxins and inhibitors; 2 = gibberellin-like substances.

taneous onset of accelerated elongation (Khrianin 1964). The bioassay shows that peak auxin activities correspond to R_f values of 0.3 to 0.4 and 0.6 to 0.8 and that peak gibberellin-like substances (GLS) have R_f values of 0.5 to 0.6 and 0.8 to 0.9. The content in natural inhibitors is relatively insignificant, and their R_f values are very high. It should be noted that a high auxin content in the leaves is matched by a relatively large quantity of gibberellins.

The histogram presented in Figure 5.1 shows that the biological activity of auxins and gibberellins is much higher in the leaves of male plants than in those of females.[1] Correspondingly, it is well known that at this particular stage male plants differ from females in their higher physiologic activity, growth rate, and accumulation of dry matter (Grishko 1935; Valter et al. 1940; Chailakhyan 1947; Senchenko et al. 1963; Djaparidze 1965; Khrianin 1967).

The results thus lead to the conclusion that the difference in the contents of endogenous growth regulators and the changes in the ratios of GLS, auxins, and

[1]Atsmon et al. (1968) also reported that shoot diffusates and root exudates from male plants have a higher GA content than those from isogenic female plants. There were several gibberellins present [Ed.].

inhibitors are clearly paralleled by, or linked to, differences in the speed and duration of growth cycles in both male and female hemp plants (Khrianin 1974, 1975, 1977).

With the data now available on the endogenous levels of growth regulators in developing hemp plants, it was possible to conduct specific experiments aimed at determining the biologic activity of the phytohormones, especially those that seem to play leading roles in sex expression in dioecious plants. The experiments of Chapter 4 strongly suggest that the role of roots in female sex expression is linked to the root-specific synthesis of cytokinins, while the role of leaves in male sex expression involves the leaf-specific synthesis of gibberellins. The next step, therefore, was to determine the endogenous levels of cytokinins and gibberellins both in the roots and in the leaves. The following experiment consists of: 1) the analytical extraction of cytokinins and gibberellins from the roots and the leaves of dioecious plants, and 2) the determination of the biological activity of these extracted phytohormones. The integral model for sex expression (Figure 4.1) was the rationale for this experimental design.

The experiment was conducted on both hemp (*cv.* US-6) and spinach (*cv.* Victoria). The plants were divided into the following sets: 1) grown in boxes containing soil, 2) grown intact in vessels, 3) grown in vessels with the adventitious roots left in place but the leaves removed, and 4) grown in vessels with leaves present but the roots removed. The roots were either allowed to regenerate (sets 2 and 3) or were systematically removed (set 4). In set 3, all the leaves, with the exception of the two apical ones, were removed after the regeneration of the adventitious roots to prevent the death of the plants.

The biological activities of cytokinins and gibberellins present in the leaves, and in the primary and adventitious roots, were determined at the beginning of sexual differentiation of the experimental plants.

In both hemp and spinach plants, the cytokinin activity was weak in the ether fraction, more significant in the aqueous fraction, and very high in the N-butanol fraction. The ether fraction was included in the experiment because several investigators have reported that ether fractions contain kinetin, zeatin, and isopentenyladenosine (Hemberg and Westlin 1973; Lethan 1974). The histograms (Figure 5.2) demonstrate that the cytokinin activity is higher in the leaves, and particularly in the primary roots, of female plants, than in the corresponding parts of male plants. This finding is entirely consistent with the earlier experiments of Engelbrecht (1973) in which the bioassay used the growth of tobacco callus.

Cytokinin activity corresponding to the R_f values of 0.7 to 0.8 is found in the leaves and roots of both male and female plants. However, a spot with cytokinin activity that has a lower mobility (R_f 0.2 to 0.3) was found in both the leaves and the roots of female plants only. The R_f values of 0.6 to 0.8 correspond either to zeatin or to kinetin markers (standard lanes in Figure 5.2). Experiments conducted on cotyledons of pumpkin had shown that the active zone with R_f values of 0.6 to 0.7 includes zeatin and its riboside, while the fraction of lower mobility (R_f 0.1 to 0.3 in the butanol-ammonia system) is of a very different nature

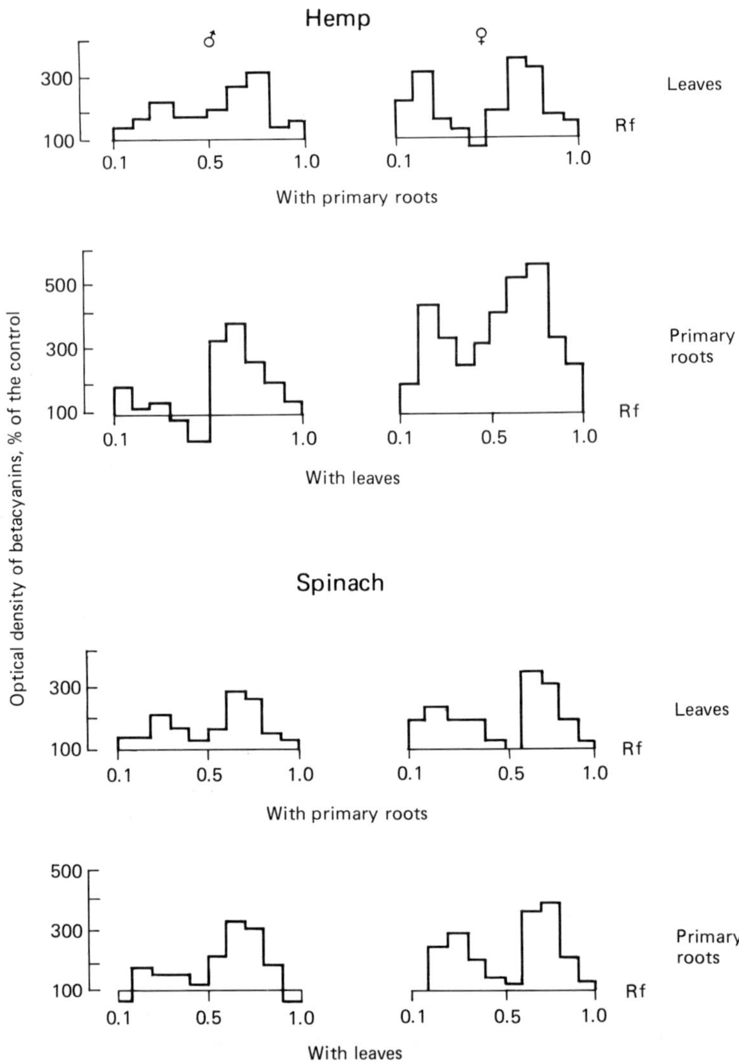

FIGURE 5.2. Biological activity of cytokinins in the leaves and the primary roots of hemp and spinach plants (grown in boxes of soil).

(Ribitska et al. 1977). Analogous results have been obtained by Henson and Wareing (1976). Thus, it is likely that in our experiment, the active species with the R_f values of 0.7 to 0.8 are zeatin and zeatin-riboside. On the other hand, little can be said about the cytokinin activity at lower mobility (R_f 0.2 to 0.3) that is apparent in the leaves of the female plants only. It may well be accounted for by derivatives of zeatin that have different physiological activities (Skoog, Hamzi, and Szweykowska 1967; Skoog and Leonard 1968; Kulaeva 1973).

Interestingly, the isolation of natural cytokinins from cucumber also yielded two cytokinin-like substances from pistillate flowers, and only one from staminate flowers. During the IXth Timiryazev lecture series in 1948, Sabinin[1] (1949) noted that the roots are the site of synthesis of hormonally active substances that are derived from nucleic acids. By now it has become well known that the initial synthesis of cytokinins occurs in the tips of primary roots and especially in the tips of lateral roots (Mothes et al. 1959; Mothes 1964; Sittin et al. 1967; Letham 1967; Wareing and Seth 1967; Münshe et al. 1968; Skene 1975; Sotta 1978). The cytokinins then are transported upward in the transpiration stream (Mothes and Engelbrecht 1961; Kulaeva 1962; Skene and Kerridge 1967).

A notably high cytokinin activity is present in the adventitious roots of intact hemp and spinach plants (set 2, Figure 5.3). It is almost twice as high as the activity in the primary roots (Figure 5.2). In sets 2 and 3, only 10% to 15% of the plants turned out to be male, and the amount of freeze-dried material was too low to determine the biologic activity of cytokinins present. Henson and Wareing (1976) have shown that the root system is capable of producing cytokinins even if the upper part of the plant is removed. This is demonstrated in our experiment in those plants that were systematically deprived of their leaves (set 3, Figure 5.3) and yet retained most of the cytokinin activity. It is quite possible that the removal of leaves interferes with the upward transport of cytokinins, which consequently are accumulated in the roots. Such a possibility has been put forward by other investigators (Sogur and Gamburg 1979). Moreover, the introduction of exogenous cytokinins can inhibit root growth (Yang Da-ping and Dodson 1970).[2]

Assuming that the leaves are normally supplied with cytokinin-like substances that originate in the roots (Mothes and Engelbrecht 1963; Kulaeva 1973), it was expected that no cytokinins would be present in the leaves of plants from which roots have been removed (set 4). Indeed, only a very weak cytokinin activity could be determined in these plants (Figure 5.3). This finding is in good agreement with the experiment of Sogur and Gamburg (1979) with pea plants, in which the removal of roots abolished the upward transport of cytokinins, and resulted in a low cytokinin content in the cotyledons and more generally in the aerial parts.

Thus, the cytokinin activities are quite different in the roots and the leaves of male and female hemp and spinach plants. In all cases, 1) the biological activity of cytokinins is higher in females, and 2) the cytokinin content is considerably higher in the roots than in the leaves. A particularly high activity is present in adventitious roots. It is evidently the regeneration of roots and their subsequent function that are responsible for the enhanced formation of female flowers in the experimental plants. The method used to determine the biological activity of

---

[1]The fact that this was 6 years before the discovery of kinetin, the first of the cytokinins, is a remarkable tribute to Sabinin's perception [Ed.].

[2]It certainly inhibits the formation and outgrowth of lateral roots (Wightman et al.: Physiol. Plantarum 49:304–314, 1980) [Ed.].

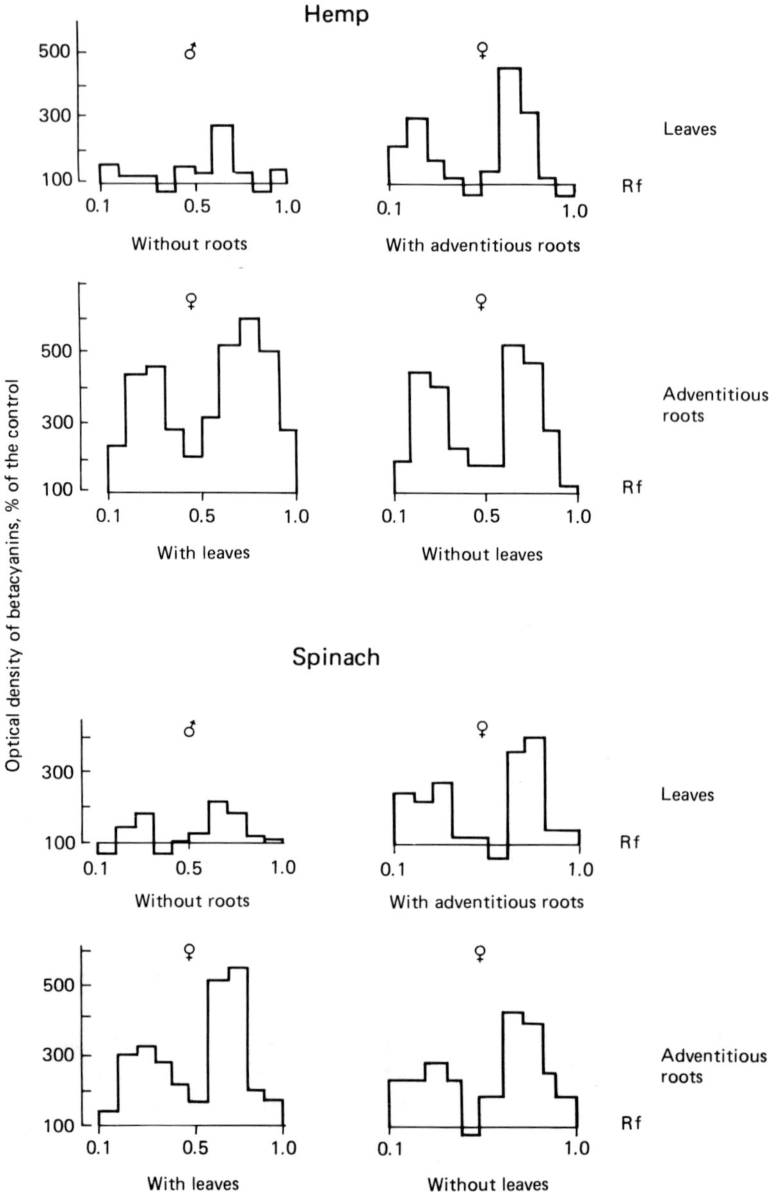

FIGURE 5.3. Biological activity of cytokinins in the leaves and the adventitious roots of experimental hemp and spinach plants grown in nutrient solutions.

gibberellins did not allow a clear determination of the presence of the phytohormone in the roots of the experimental plants. Gibberellin activity is known to be present in the roots of plants such as *Vitis vinifera* L., *Oryza sativa*, *Pharbitis nil*, and *Ipomoea batatas*, among others (Skene 1967; Muvakami 1968). However, gibberellins are also synthesized independently in the leaves (Chailakhyan 1958, 1968; Stoddart and Lang 1967) and are transported to the roots later, where they are modified (Crozier 1970). In the present experiments, the analysis of gibberellins present in the leaves during sexual differentiation (beginning of bud development) revealed that their activity is higher in male hemp and spinach plants than in the females. Similar results have been obtained earlier (Khrianin 1975, 1977). Chromatographic experiments conducted on cucumber demonstrated earlier than GLS activity is considerably higher in monoecious plants than in dioecious ones (Atsmon et al. 1968). The presence of adventitious roots (set 2) results in weak gibberellin activity in the leaves (Figure 5.4). As was mentioned earlier, the leaves and roots of the plants of this set have a characteristically high cytokinin activity. We conclude that it is indeed the shift in the ratio of endogenous phytohormones in favor of cytokinins that results in enhanced female sex expression in these plants. The biological activity of gibberellins present in the leaves of plants deprived of their roots (set 4) is sufficiently high (Figure 5.4) that, despite the presence of moderate amounts of cytokinins, male sex expression is nonetheless favored.

The results of the present experiment demonstrate inter alia that the levels of cytokinins and gibberellins are not independent variables. High cytokinin activity coincides with low gibberellin activity, and *vice versa*. This finding reinforces the notion that roots, being cytokinin producers, play an important role in leaf metabolism, and that leaves, being producers of gibberellins, are involved in the regulation of root metabolism (Kulaeva 1973; Chung and Chung 1978). Furthermore, the results obtained are most valuable in understanding the physiological significance of growth regulators in plant morphogenesis and flowering (Chailakhyan 1975; Krekule and Seidlova 1977; Miginiac 1978; Sotta 1978).

It should be noted that the removal of the tips of the root system leads to early leaf senescence and decreased rates of CO_2 fixation (McDavid et al. 1972). This effect can be avoided if the plants are treated with synthetic cytokinins. Alternatively, the removal of leaves leads to root senescence, while gibberellin treatment inhibits root formation (Gaspar et al. 1977). These findings imply that there are specific interactions between gibberellins and cytokinins that participate in essential biochemical and physiological processes, of which sexual differentiation is only one example.

Thus, the regulation of sex expression involves an interplay between different organs that are sites of synthesis of phytohormones. Knowledge of the levels of endogenous cytokinins and gibberellins directly confirms the earlier hypothesis that roots behave like enhancers of female sex expression because of the root-specific synthesis of cytokinins, whereas leaves behave like enhancers of male

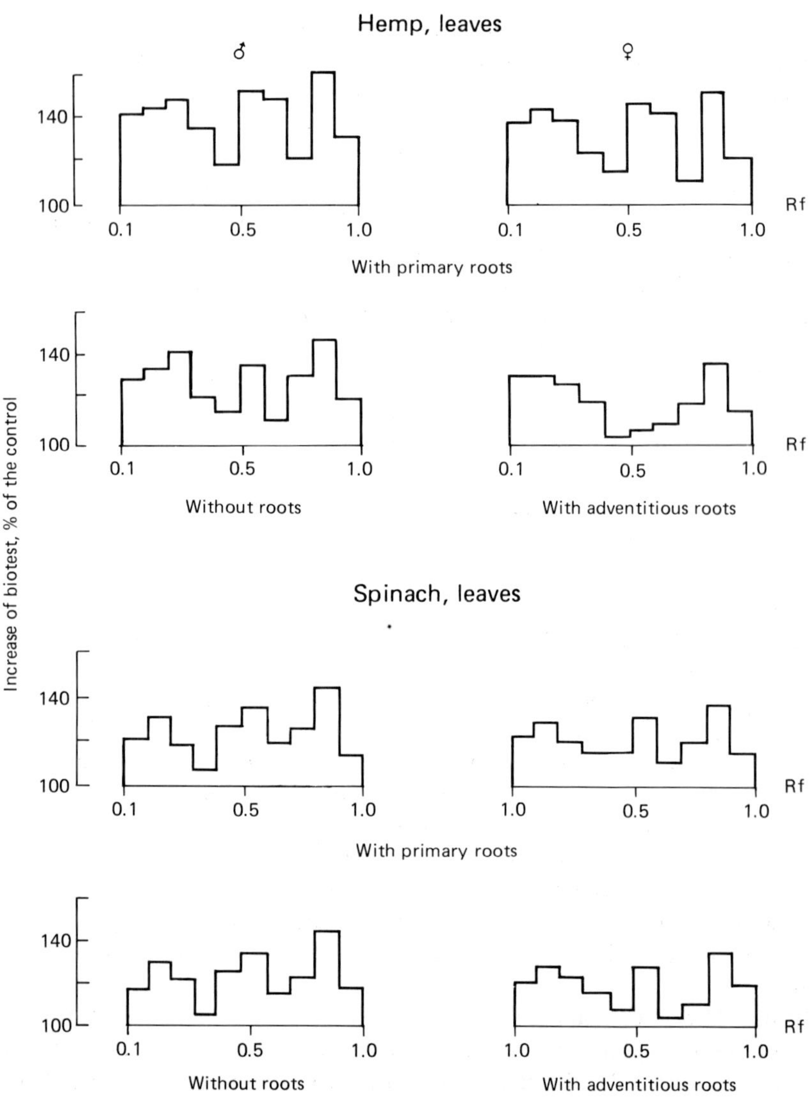

FIGURE 5.4. Biological activity of gibberellins in the leaves and roots of intact and mutilated hemp and spinach plants.

sex expression because of the leaf-specific synthesis of gibberellins. The data obtained confirm the predictions of Sabinin (1963, 1971) that the meristems of plant organs must produce specific substances that determine the type of sex expression.[1] Moreover, the results explain why growth in short days leads to dioecious species to an increase in the number of female individuals, while long days favor the formation of male individuals (Schaffner 1919; McPhee 1924; Valter and Lilienshtern 1934; Khrianin and Chailakhyan 1977). It has been determined that exposure of hemp plants to a 10-hour photoperiod for 6 to 12 days results in enhanced development of the root system and in a decrease in the growth in height (Makarevich 1935; Levchenko 1937). There is no doubt that such a strong change in the ratio between the growth of roots and that of the rest of the plant shifts the balance of phytohormones in favor of cytokinins, which, of course, results in enhanced female sex expression. On the other hand, long days increase the growth rate and the mass of the aerial part of the plant (Makarevich 1935; Levchenko 1937). Consequently, the level of gibberellins is increased and so is the proportion of male individuals. Male plants tend to have more developed aerial parts and female plants have a more developed root system, as shown in section B of Chapter 4 (Dobrunov 1935; Molotkovskyi 1976).

Thus, changes in the photoperiod alter the proportion of roots to shoots, which in turn differentially affects the content of phytohormones, in particular gibberellins and cytokinins. The result is the appearance of sexual characteristics of the male or female type.

---

[1]In this respect, present day plant physiology comes curiously close to the conception of Julius Sachs a century ago. His "root-forming substance" can be identified with auxin and his "stem-forming substance" with gibberellins. However, his "leaf-forming substance" does not come close to cytokinin, at least as yet. Sachs, however, made no such statement about sex expression [Ed.].

6 Hormonal and Genetic Factors in Plant Sex Expression

A. Sex Expression in Isolated Hemp Embryos Grown in Culture

The direct effect of phytohormones on sexual differentiation was studied in a culture of isolated embryos of hemp (*Cannabis sativa*, cultivar US-6).

The seeds were surface-sterilized in a weak solution of $KMnO_4$ or in 0.1% $MgCl_2$ for 15 and 5 minutes, respectively, then washed several times in sterile water and transferred to 5 to 6 layers of damp filter paper and placed in Petri dishes. The rigid bivalvular seed coats burst after 14 to 18 hours. The mature embryos were removed from the seeds and the cotyledons cut off. The embryos were transferred into sterile test tubes containing White's agar which includes minerals, vitamins, and sucrose (White 1949; Butenko 1964). The growth conditions included lighting, 0.002 W/cm^2; temperature, 20 to 22°C; and relative humidity, 80%. For the first 7 days the photoperiod was 16 hours, and thereafter 8 hours, until sex expression became apparent.

Three experiments were carried out. Each experiment included 3 sets: 1) pure White's medium, control; 2) GA, 12.5 mg/L; 3) 6-BAP, 5 mg/L. The optimal levels of these substances had been determined in preliminary experiments. Each set contained 30 to 50 embryos. During the course of the experiment, 1 to 5 embryos died in each set. The experiments were started on June 22, August 3, and August 30. The results are presented in Table 6.1. Even at the early stages of the experiment it became apparent that the growth rates varied considerably from one set to another (Table 6.1 and Figure 6.1); these differences were maintained throughout each experiment. The GA-treated plants began budding 10 days earlier than the control plants, while 6-BAP was slightly inhibitory.

In addition to the effects on growth and development, the phytohormones strongly influenced sex expression. In all 3 experiments, the introduction of GA into the nutrient medium resulted in the almost exclusive formation of male plants (95.5% to 100%), while 6-BAP induced female sex expression (92.6% to 97.7%, Table 6.1 and Figure 6.2). In other words, the effects of GA and 6-BAP are much more drastic when these phytohormones are applied to embryos than when introduced through the sectioned stems or the roots of young seedlings.

TABLE 6.1. Effects of gibberellins and cytokinins on growth, development, and sex expression in isolated hemp embryos (combined results of three experiments).

Treatment	Height of plants (cm) June 28	August 28 Male plants	August 28 Female plants	Beginning of flower bud formation (July)	Male plants No.	Male plants %	Female plants No.	Female plants %
Control	2.5	4.5	3.5	25	12	42.9	16	57.1
					26	54.6	22	45.4
					20	42.6	27	57.4
				Average		46.7		53.3
GA	4.8	7.4		15	30	100	0	0
					46	97.9	1	2.1
					43	95.5	2	4.5
				Average		97.8		2.2
6-BAP	1.7		2.3	27	2	7.4	25	92.6
					1	2.3	44	97.7
					2	4.2	46	95.8
				Average		4.6		95.4

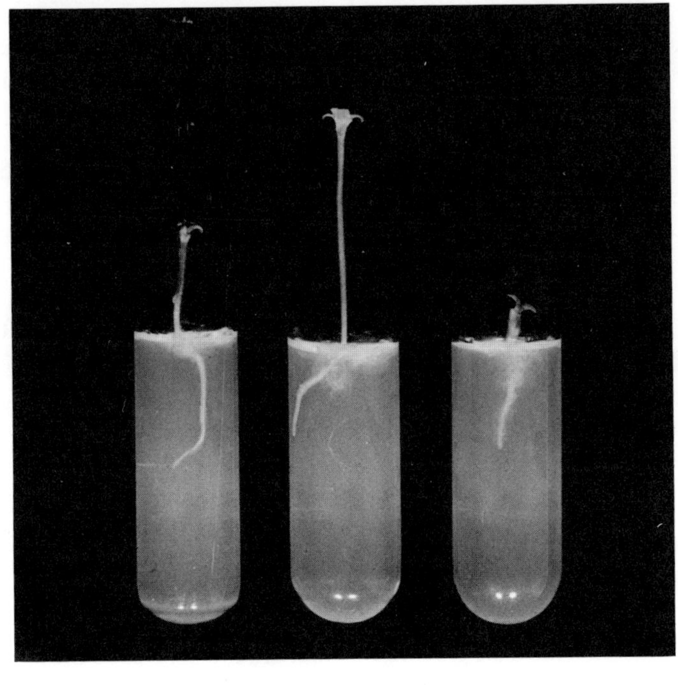

FIGURE 6.1. Hemp seedlings in 1 = White's medium; 2 = White's medium plus GA; 3 = White's medium plus 6-BAP.

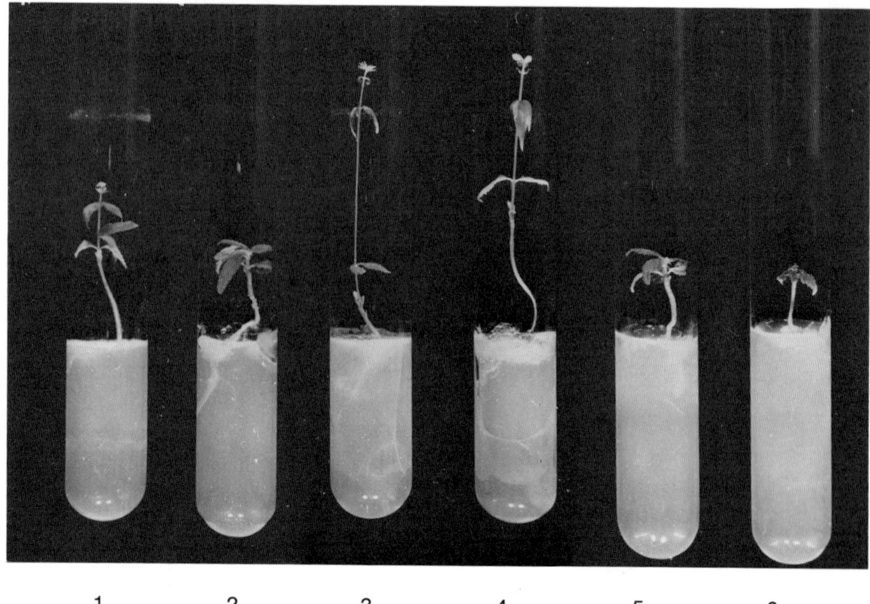

FIGURE 6.2. Sex expression in hemp plants. 1 = White's medium only, a male plant; 2 = White's medium only, a female plant; 3 and 4 = White's medium plus GA, male plants; 5 and 6 = White's medium plus 6-BAP, female plants.

This finding implies that the culture of isolated embryos is likely to be the method of choice in studying the hormonal regulation of sex expression in dioecious plants, and, possibly, in monoecious plants carrying unisexual flowers.

B. Generalized Concept of Sex Expression in Dioecious and Monoecious Plants

The experiments on the effects of introducing exogenous phytohormones into plants have yielded results that are entirely consistent with the subsequent analytical studies on the variations in the levels of the same phytohormones already present. The consistent result in the various experiments adds validity to our general ideas regarding the shift in the directionality of sexual differentiation in dioecious plants, namely, in hemp and spinach. Based on our experimental work, we can now draw a general picture of sex expression in dioecious plants (Figure 6.3). The leaves synthesize gibberellins that are transported to the apical buds of the shoots, where they induce a number of changes leading to male sex expression (Figure 6.3, no. 1). On the other hand, the roots produce cytokinins that are also transported to the apical buds, where they tend to induce female sex formation (Figure 6.3, no. 4). Under favorable natural conditions, the balance of the phyto-

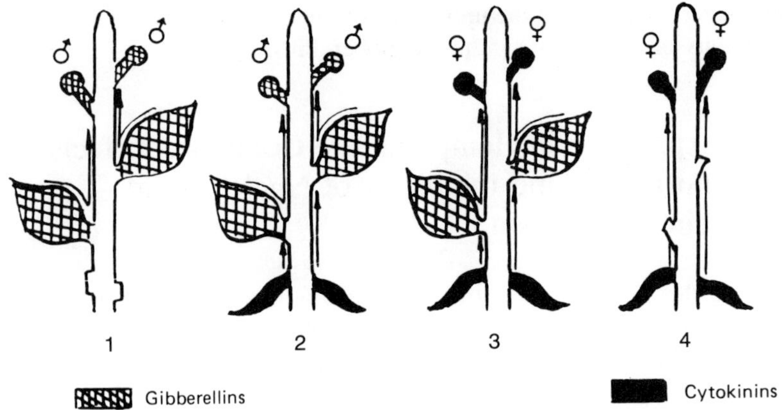

FIGURE 6.3. Roles of roots, leaves, cytokinins, and gibberellins in sex expression of dioecious plants (see text).

hormones is such that sex expression appears to be exclusively under the control of the genetic apparatus, which explains the formation of an equal number of male and female plants (Figure 6.3, no. 2 and 3).

The mechanisms of action of phytohormones in monoecious plants carrying unisexual flowers are slightly more complex because male and female flowers are carried on the same plant. In cucumber, for example, in the first stage of development, only male flowers are produced; in the second stage, the flowers are mixed, while in the third, most of the flowers are female (Frankel and Galun 1977).[1] This developmental pattern can be ascribed to the relative sizes of the root system and of the leaf area. Initially, the leaf area is more developed than the root system and the plant is accumulating gibberellins that are responsible for the formation of male flowers. In the second stage, both the root system and the leaf area are well developed, which ensures the production of both gibberellins and cytokinins. Therefore, male and female flowers are likely to be formed. In the third and last stage of development, there is a gradual decrease in the activity of the leaves with a still functionally intact root system. Hence, the phytohormones that are transported to the apical buds are predominantly cytokinins, which of course, accounts for an increase in the number of female flowers.

In accordance with the ideas of Molotkovskyi on sex expression in corn (1960, 1968), the formation of female flower clusters (ears) in the lower part of the plant is entirely logical, since the transport of cytokinins to an area adjacent to the root system is no doubt very easy. Following the same logic, the formation of male

[1]There are four stages in the acorn squash (Nitsch et al.: Am. J. Bot. 39:32–43, 1952): 1) male flowers only, 2) perfect flowers, 3) female flowers, and 4) "superfemales," which produce pathenocarpic fruit. In this plant, the trend of increasing femaleness was related to a strong trend of increasing auxin content [Ed.].

flower clusters (anthers) in the upper part of the corn plant can be explained by the proximity of leaves that synthesize gibberellins.[1]

C. Sex Expression in Hemp and the Combined Effects of Phytohormones and Inhibitors of Nucleic Acid and Protein Metabolism

In studying the role of phytohormones in plants, many researchers have been interested in possible stimulatory effects that phytohormones may have on the induction of the synthesis of nucleic acids and proteins (Merkis et al. 1971; Kulaeva 1973, 1975; Muromtsev and Agnistikova 1973; Kefeli 1973). The present belief is that phytohormones act by regulating gene activation within cells (Bonner 1967, 1968; Khrianin 1969, 1978; Gamburg 1976; Kulaeva 1977; Aleshin et al. 1978). A common experimental approach to this problem is the use of inhibitors that act to varying degrees on the synthesis of nucleic acids and proteins.

We have used this approach in the hope of gaining some understanding of the regulatory mechanisms of phytohormones in sex expression in hemp. Our experiments on the induction of male sex expression by gibberellin, as well as independent reports indicating that gibberellins stimulate RNA and protein synthesis (Muromtsev and Agnistikova 1973) led to the idea that the gibberellin promotion of male sex expression would be eliminated if the plants were treated with an inhibitor of RNA synthesis, such as actinomycin D. An experiment aimed at testing this idea follows.

Hemp plants (*cv.* US-6) were grown in boxes of soil in 18-hour days.[2] At the stage of 2 to 3 leaf-pairs, the plants were sprayed with a solution of gibberellin A_3 (GA_3), 25 mg/L and/or actinomycin D (Act. D) at a concentration of 1 mg/L. The optimal concentrations had been determined in preliminary experiments. The treatment was repeated 24 hours later. The experiment included the following sets: 1) control, treated with water; 2) Act. D; 3) Act. D, and 24 hours later, GA; 4) GA, and 24 hours later, Act. D; and 5) GA. Each set included 100 plants. The experiment was repeated twice.

As expected, the treatment of hemp plants with GA stimulated their growth in height; the plants treated with Act. D, on the other hand, were smaller than the controls (Figure 6.4, no. 2). Spraying with Act. D before or after the GA treatment (Figure 6.4, no. 3 and 4) partially counteracted the growth stimulation due to GA. At the end of the experiment, the heights of the experimental plants were

[1]It might be difficult to apply this argument to the aroids, in which it is true that the female flowers are usually at or near the base of the spadix and thus nearer the roots. However, they are also nearer the attachment of the spathe, which represents most of the green tissue [Ed.].

[2]Location: Timiryazev Plant Physiology Institute.

FIGURE 6.4. Effects of GA and actinomycin D (Act. D) on growth and sex expression in hemp plants. 1 = Control; 2 = Act. D; 3 = Act. D plus GA; 4 = GA plus Act. D; 5 = GA.

as follows: 1) control, 92 cm (males) and 54 cm (females); 2) Act. D, 69 and 46 cm; 3) Act. D plus GA, 116 and 100 cm; 4) GA plus Act. D, 117 and 102 cm; 5) GA, 156 and 147 cm. Act. D also retarded the flowering, which was delayed by 5 days compared with the other sets.

The distribution of sexes was considerably affected by the treatments (Table 6.2). As expected, GA treatment resulted in an increase in the number of male plants. The treatment with Act. D, on the other hand, led to a slight increase in the number of female plants. The combined treatments Act. D plus GA and GA plus Act. D demonstrated that the inhibitor of RNA synthesis counteracts the effect of GA in inducing male sex expression in hemp.

In the experiments of Jaiswal and Mohan Ram (1974), the simultaneous treatment of hemp plants with GA (25 mg per plant) and cycloheximide (10 to 25 mg per plant), an inhibitor of protein synthesis, resulted in a significant decrease in the number of nodes giving rise to male flowers. The growth was also slightly

TABLE 6.2. Effects of GA and actinomycin D (Act. D) on sex expression in hemp plants.

Treatment	Percentage of plants		
	Male	Female	Monoecious females
Control	54.5	45.5	0
GA	72.7	0	27.3
Act. D	38.8	61.2	0
Act. D + GA	47.2	52.8	0
GA + Act. D	54.0	46.0	0

inhibited. When the concentration of cycloheximide was raised to 50 mg per plant, practically no male flowers were formed, while general growth was significantly impaired. Jaiswal and Mohan Ram proposed that this effect of cycloheximide was due to the inhibition of an enzymatic activity that ultimately results in the appearance of male flowers.

Considering that in hemp the male genetic program is realized somewhat earlier than the female program, the effect of Act. D can be thought of as the inhibition of the more active male developmental program that is stimulated by gibberellins. Alternatively, Act. D can be seen as an enhancer of female sex expression; by inhibiting the male developmental program, it allows the activation of the female program. Thus, the mode of action of GA on sex expression is related to the realization of a genetic program. GA is apparently a modulator of the processes responsible for male sex expression in hemp.

Further experiments were aimed at determining the individual and combined effects of GA, 6-BAP, inhibitors of replication (mitomycin), or transcription (Act. D), and of translation (puromycin). The integral model of sex expression (Chapter 4) was the rationale for the experimental design. It is well known that the antibiotics used, in addition to their specific targets, i.e., DNA, RNA, and protein synthesis, have occasional and undesirable side effects. The effective concentrations of the antibiotics were therefore kept at a minimum.

The experiment was carried out essentially as before. Hemp plants (*cv.* US-6) were pregrown in boxes of soil in 18-hour days. At the stage of three leaf-pairs, the plants were cut at the level of the root collar and transferred to vessels containing either water (for the control) or a solution of phytohormones and/or inhibitors. The plants were then allowed to grow in a growth chamber in 16-hour days, at 20°C and 80% humidity. The experimental plants were treated with the test substances for a period of 28 hours. The combined treatments were staggered in time, treating the plants first with the phytohormones and then with the inhibitors. After treatment, the plants were transferred to Knop's solution.

The experiment included the following sets: 1) control; 2) mitomycin, 1 mg/L; 3) Act. D, 1 mg/L; 4) puromycin, 1 mg/L; 5) GA, 25 mg/L; 6) GA, 25 mg/L plus mitomycin, 0.5 mg/L; 7) GA, 25 mg/L plus Act. D, 0.5 mg/L; 8) GA, 25 mg/L plus puromycin, 0.5 mg/L; 9) 6-BAP, 15 mg/L; 10) 6-BAP, 10 mg/L plus mitomycin, 0.5 mg/L; 11) 6-BAP, 10 mg/L plus Act. D, 0.5 mg/L; 12) 6-BAP, 10 mg/L plus puromycin, 0.5 mg/L. After the regeneration of adventitious roots, the leaves of the plants in sets 5 through 8 were removed. In sets 9 through 12, the roots were systematically removed. Each set included 112 plants, and the experiment was repeated 8 times.

Among all the inhibitors, only Act. D slightly inhibited the growth in height of male and female hemp plants. The stimulatory effect of GA on the growth of both the male and female plants was inhibited by only Act. D; all the other inhibitors counteracted the stimulatory effect of GA in male plants only. The introduction of 6-BAP (in the concentrations used) also inhibited the growth. This inhibitory effect was slightly enhanced by the additional treatment with mitomycin or puromycin. Act. D, on the other hand, did not alter the growth inhibition caused by 6-BAP.

The inhibitor alone did not cause any delays in bud development. Buds on the plants treated with GA began opening 4 days earlier than on the control plants. Of the three inhibitors, only Act. D prevented the early bud development caused by GA. The treatment with 6-BAP and 6-BAP plus Act. D delayed bud development by 2 days, while the additional treatment with mitomycin or puromycin caused a total delay of 5 days. The detailed results of the experiment are presented in Table 6.3 and Figures 6.5 and 6.6.

As in the earlier experiments, female sex expression was favored in the control plants (that carried leaves and newly regenerated roots). The sex ratio in the plants treated with Act. D was comparable to the sex ratio in the controls. The treatments with mitomycin and puromycin, however, resulted in an increase in the number of male plants.

As expected, the GA treatment led to an increase in the number of male plants. This effect of GA was counteracted by Act. D only; mitomycin and puromycin were ineffective (Figure 6.5). We tentatively conclude that the stimulatory effect of GA on male sex expression and RNA synthesis are somehow connected.

The introduction of 6-BAP resulted in an increase in the number of female plants. Both mitomycin and puromycin interfered with this effect of 6-BAP; Act. D was ineffective (Figure 6.6). Because mitomycin inhibits DNA replication and puromycin stops protein synthesis, it follows that the mode of action of 6-BAP in promoting female sex expression involves both DNA and protein synthesis.

It must be admitted that the results of this experiment constitute only very indirect evidence for the participation of replication, transcription, and translation in the processes of sex expression. Much stronger evidence would be

TABLE 6.3. Effects of phytohormones and nucleic acid and protein inhibitors on growth, development, and sex expression in hemp plants.

Treatment	Final height of plants (cm)		Beginning of flower bud formation	Percentage of plants		No. of male flowers per male plant
	Male	Female		Male	Female	
Control	42.3	33.0	May 4	26.3	73.7	43.0
Mitomycin	41.2	32.8	May 4	65.7	34.3	46.7
Act. D	35.6	30.8	May 4	37.2	62.8	34.5
Puromycin	40.8	32.3	May 4	56.3	43.7	43.3
GA	50.7	40.5	April 30	80.9	19.1	49.6
GA+mitomycin	42.5	39.0	April 30	66.6	33.4	47.3
GA+Act. D	36.0	32.6	May 4	28.5	71.5	39.0
GA+puromycin	40.3	38.0	April 30	65.2	34.8	45.7
6-BAP	36.0	31.4	May 6	23.3	76.7	31.5
6-BAP+mitomycin	30.0	27.2	May 9	72.7	27.3	35.6
6-BAP+Act. D	35.6	31.0	May 6	22.2	77.8	30.0
6-BAP+puromycin	28.3	26.1	May 9	75.8	24.2	36.3

FIGURE 6.5. Effects of GA and inhibitors of nucleic acid and protein synthesis on sex expression in hemp plants. 1 = GA; 2 = GA plus mitomycin; 3 = GA plus actinomycin D; 4 = GA plus puromycin.

provided by direct analytical studies. Nevertheless, it appears likely that the mode of action of phytohormones in determining the type of sex expression in hemp involves changes in the activity of the genetic apparatus. GA may well act at the level of transcription, whereas 6-BAP could act at the level of replication and translation (see diagram, Figure 6.7).

It is necessary to emphasize that there is no direct proof that the phytohormones induce protein synthesis by regulating gene expression (Kulaeva 1977).[1]

[1]There is, of course, plenty of evidence that auxin action involves modification in the formation of both RNA and proteins. There is some evidence of the same kind for GA and cytokinin action (see, for example, the review of Jacobsen: Ann. Rev. Plant Physiol. 28:537–564, 1977) [Ed.].

3

4

FIGURE 6.5. *Continued.*

Such a mode of action is considered proven in the case of animal steroid hormones and specific genes in target tissues. For example, Woo and O'Malley (1976) have demonstrated that estrogen induces the synthesis of ovalbumin in chick oviducts by inducing the synthesis of ovalbumin RNA. The changes they observed in the chromatin matrix apparently reflect such an activation of the corresponding genes. Steroid hormones have a common mode of action: they alter the functional state of the genetic apparatus of the cell, making possible subsequent chemical reactions that result in gene activation (Sergeev et al. 1971). Steroid hormones, like all the other animal hormones, act on gene expression by forming hormone-receptor complexes (Staroseltseva 1976; Kulaeva 1977). Hormonal receptors, which are usually proteins, are found in the nucleus, cytoplasm, lysosome, and in the membranes. Certain phytohormones, particularly gibberellins, have a chemical structure that somewhat resembles the structure of steroid hormones (Heftmann 1972; Muromtsev and Agnistikova 1973; Paseshnichenko

FIGURE 6.6. Effects of 6-BAP and inhibitors of nucleic acid and protein synthesis on sex expression in hemp plants. 1 = 6-BAP; 2 = 6-BAP plus mitomycin; 3 = 6-BAP plus actinomycin D; 4 = 6-BAP plus puromycin.

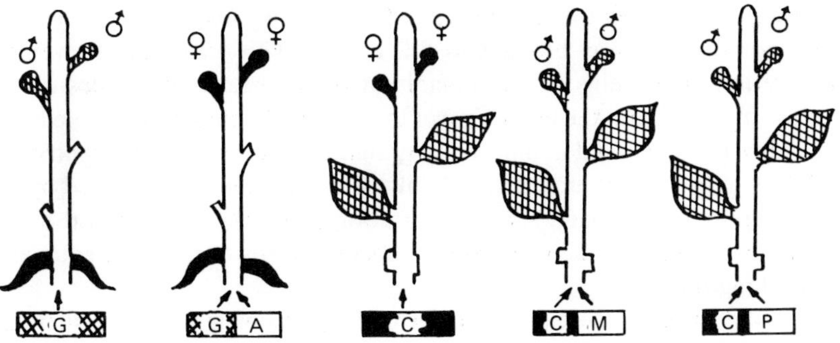

FIGURE 6.7. Simultaneous action of phytohormones and inhibitors of NA and protein synthesis on sex expression in hemp. G = gibberellins; C = cytokinins; A = actinomycin D; M = mitomycin; P = puromycin.

and Guseva 1974). It is probable, therefore, that the phytohormones also have specific receptors that regulate gene expression. Indeed, although there is no direct evidence as to the nature of such receptors (Kende and Gardner 1976), there are indications of interactions between auxins and proteins (Hertel et al. 1972; Kefeli 1973), gibberellins and proteins (Muromtsev and Agnistikova 1973; Higgins et al. 1976), and cytokinins and proteins (Takegami and Yoshida 1975; Kulaeva 1977).

D. Immunochemical Analysis of Stem Apices of Male and Female Hemp Plants

The use of inhibitors of nucleic acid and protein metabolism did give some indications as to the possible modes of action of different phytohormones. Here we present another approach aimed at determining the regulation of sex expression at the level of genes. In essence, it consists in comparing the sets of proteins present in male and female hemp plants. As was found in *Mercurialis annua* L., for example, male and female plants have different sets of antigens (Durand-Rivières 1969).

Hemp plants (*cv.* US-6) were grown either in soil under natural conditions or in boxes of soil in 18-hour days. As soon as sex expression became apparent, the stem apices were cut off. The degree of differentiation of flower primordia was estimated using light microscopy. The collected plant material was fixed in liquid nitrogen and freeze-dried.

The analysis was based on the immunologic procedure initially developed by Zilber and Abelev (1962) and modified for plant tissues by Volodarskyi (1971, 1973). The freeze-dried plant tissue was finely ground and resuspended in physiological saline (0.14 M NaCl). The suspension was stirred at 4°C for 18 hours. The volume of saline added depended on the protein content in the plant material, which was determined by the method of Lubochinsky and Zalta (1954). The pH of the homogenate was brought to 8.2 to 8.6. This treatment prevented the denaturation of proteins by the organic acids, and favored the destruction of cellular structures and the extraction of the maximal number of soluble antigens. Tissue debris, intact nuclei, and starch granules were removed by centrifugation at 4000 $\times$ *g* for 15 minutes. The solubilized antigens were preserved by adding the antiseptic, merthiolate, at a concentration 1:10,000. The pH of the solution was then adjusted to 7.2. The protein concentrations of the final extracts were 5 and 10 mg/mL.

When the antigens were prepared for rabbit immunization, the centrifugation step was omitted. The homogenate was kept in the refrigerator for 1 to 2 hours and then injected into the rabbits (Chinchilla variety). The antisera were raised against tissue homogenates of hemp stem apices using standard techniques (Zilber and Abelev 1962) with the addition of complete Freund's adjuvant and antibiotics.

The first injection was performed hypodermically in the region of the stomach. The solution injected contained tissue homogenate (10 mg protein), adjuvant, 100,000 U penicillin and 100,000 U streptomycin. The following proportions were used: 1 vol antigen plus 1 vol lanoline plus 2 vol vaseline plus 2.5 mL BCG in vaseline. The second injection was performed intramuscularly 14 days after the first. The same quantity of homogenate (10 mg protein) was used but the adjuvant was omitted. The third intramuscular injection (15 mg protein) was performed 7 days later. A stimulating phlebotomy on the ear vein was performed a further 7 days later. Subsequently, the rabbits were repeatedly reimmunized once a month with the same amount of homogenate. Reimmunization led to better production of antibodies, resulting in a higher titer of the serum. The complete procedure included 5 to 7 reimmunizations of 2 to 3 rabbits (180 to 200 mg protein) for each type of plant tissue homogenate. Depending on the type of tissue and on the number of reimmunizations, the sera used had a titer of 1:32 or 1:64.

To conduct studies of individual antigens it is necessary to isolate antibodies that are specific for an antigen or group of antigens representative of a particular tissue. The highly specific antibodies were isolated by the method of Volodarskyi (1971). The antisera were prepared as described by Abelev and Avenirova (1960).

At the basis of every immunodiffusion method lies the antigen-antibody reaction; this results in the formation of a stable complex that precipitates. If the appropriate ratio is obtained, the combination of the antigen and the antibody forms sharp precipitation lines. Moreover, the number of lines corresponds to the number of different antigen-antibody pairs. There are several immunodiffusion techniques available (Ouchterlony 1958; Oudin 1960; Zilber 1968; Maurer 1971). The antigen structure of stem apices of male and female hemp plants was analyzed both by immunodiffusion and immunoelectrophoresis (Kovaleva et al. 1980). A general view of the antigen structure of apices of male and female plants is presented in Figure 6.8. The number of precipitation lines is especially indicative of this structure (Figure 6.8C,D). More detailed results were obtained with

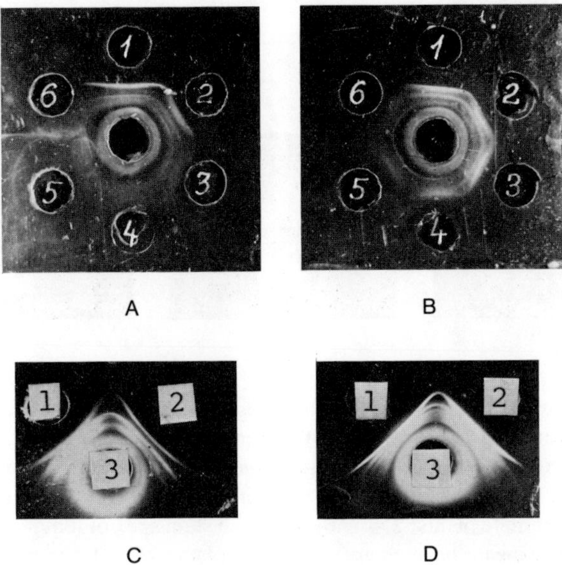

FIGURE 6.8. Protein spectra of the apices of male and female hemp plants. (A) The central well contains antiserum against proteins of the male apex; 1–6 = antigens present in the apices of male plants (1:2 to 1:32 dilution). (B) The central well contains antiserum against proteins of the female apex; 1–6 = antigens present in the apices of female plants (1:2 to 1:32 dilution). (C) 1–2 = Antigens present in the apices of male plants. (D) 1–2 = Antigens present in the apices of female plants; 3 = antiserum against proteins of the female apex.

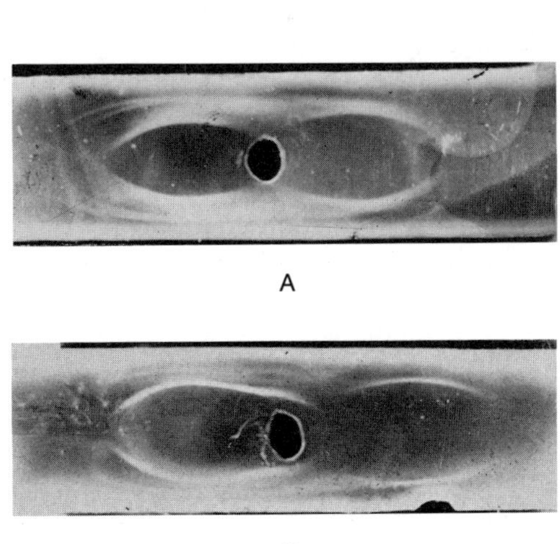

FIGURE 6.9. Immunophoregrams of antigens present in female (A) and male (B) stem apices of hemp plants.

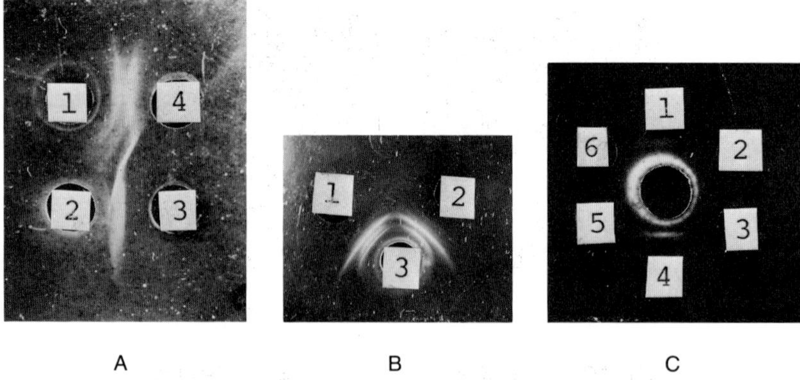

FIGURE 6.10. Comparative analysis of the antigen spectra of stem apices of male and female hemp plants. (A) 1 = Antigens in the stem apex of female plants; 2 = antigens in the stem apex of male plants; 3 = antiserum against proteins in the stem apex of female plants; 4 = antiserum against proteins in the stem apex of male plants. (B) 1 = Antigens in the stem apex of male plants; 2 = antigens in the stem apex of female plants; 3 = antiserum against proteins in the stem apex of female plants. (C) The central well contains a highly specific antiserum against proteins in the stem of female plants (saturation with antigens from the stem of male plants); 1 and 4 = antigens in the apex of female plants; 2,3,5,6 = antigens in the apex of male plants.

immunoelectrophoresis in veronal-medinal buffer (Figure 6.9). The antigen components of male and female plants differ in their electrochemical properties, diffusion coefficients, and serological specificities. Comparative analysis of the "square" method (Abelev and Avenirova 1960) revealed the specific antigens in each system (Figure 6.10). The majority of antigens are comon to both male and female plants (vertical lines in Figure 6.10A). However, the experiment revealed one antigen that is specific for the apices of female plants (the diagonal line in Figure 6.10A). This specific protein can also be seen in a slightly different experiment (Figure 6.10B).

The experiment did not reveal any specific proteins that are characteristic for the apices of male plants. Nevertheless, the analysis demonstrated that even at an early age of differentiation, apices of male and female plants differ in their protein spectra. The sera obtained were highly specific against the tissues of male and female apices. The experiment in Figure 6.10C shows a highly specific reaction between the serum against female plant tissues and the antigens present in the apices of male and female plants. The precipitation lines adjacent to wells 1 and 4 are due to a protein that is specific for the stem apices of female plants. An identical experiment using serum against male plant tissues did not reveal any proteins specific for male plants.

Thus the analyses demonstrate that at the beginning stage of sexual differentiation, the apices of female plants contain a specific protein that is absent in male apex tissues. Considering that our experiments on the biological activity of phytohormones have shown that cytokinins characteristically lead to female plants, it is quite possible that the newly discovered protein and the specific cytokinins form a hormone-receptor complex that regulates the expression of the genes responsible for female sex expression.

Suggestively, it was shown some time ago that the extraction of proteins from the meristems of female mercury plants treated with kinetin (6-furfurylaminopurine) included a specific antigen that was absent in male plants (Durand and Durand-Rivières 1969). Moreover, the tRNA-s and corresponding synthetases are more active in female mercury plants than in male plants (Bazin et al. 1975). The structural gene encoding the antigen specific for female plants is believed to be located on one of the autosomes (Durand-Rivières 1969; Durand and Durand-Rivières 1969). Thus it appears that nucleic acid and protein metabolism is more active in female plants than in males and that this higher activity is the result of gene activation.

7 Plant Sex Expression and the Interactions Between Ecologic Factors, Phytohormones, and the Genetic Apparatus

Studies of sexuality in plants have given rise to numerous theories of sex determination (Rozanova 1935). Three of these are of major importance: 1) the endogenous theory, which relates sex determination to the genetic complement, 2) the exogenous theory, which relates sex determination to the influence of external factors, and 3) the endoexogenous theory, which emphasizes both external and internal factors.

The endogenous theory was very much favored by Correns (1906). Indeed, he believed that changes in the genetic material were of paramount importance in determining sex expression. In his opinion, the genotypic and the phenotypic determination of sexuality were different and unrelated processes.

Schaffner (1927, 1935) gathered a considerable body of evidence in favor of the exogenous theory. His conclusion was simple: the genetics was completely unrelated to sex determination in plants. Instead, every plant had the potential to express either type of sexuality, which implied that sex transformation was, at least in theory, always possible. In his opinion, environmental factors exerted their influence on sex expression by altering some metabolic processes in the cells. In other words, the physical and chemical state of the plasma of the cell was essential to allow the environmentally induced changes to take place. A similar idea was put forth by Grishko (1935), who believed that the environment acted as a reagent that triggered sex transformation.

At the present time, the generally accepted theory of sex determination is the endoexogenous theory, according to which the mechanisms of sex determination and sex expression have both internal and external components. Sex determination in plants has a genetic basis, i.e., sex differentiation and sex expression are dependent on the genotype, but they are also influenced by environmental factors. Of course, as the plant develops, it is understood that the effects of a given external factor vary. The action of environmental factors is known to affect many physiological and biochemical characteristics that are specific for one sex or the other (e.g., Minina and Kushnirenko 1949; Galun 1961; Atsmon and Galun 1962; Saito and Ito 1961).

Joyet-Lavergne (1931) proposed that the influence exerted by environmental factors on the type of sex expression is accompanied by changes in the redox system of the cells. In consequence, providing that certain metabolic pathways are present, the plant forms hormonally active compounds (Sabinin 1937; Minina and Tylkina 1947). The plant's reaction to the environmental factors is due to the specific sensitivity of meristematic tissues of the apex at a particular developmental stage (Sabinin 1940; Lvova 1963a; Heslop-Harrison 1972; Khrianin and Milyaeva 1977). Recognition of this developmental stage is prerequisite for successful treatment with any factor that is known to affect sex expression. Recently, the influence of certain factors, such as the photoperiod, has been connected to the action of growth regulators that are responsible for the appearance on the apex of male or female flower primordia (Minina and Larionova 1979). According to Molotkovskyi (1968), external factors can disrupt the established interactions between the roots and the leaves, which in turn affect sex expression. Under normal conditions, the roots and the leaves of dioecious plants each influence sex expression to the same extent, so that the sex ratio is balanced.

With the discovery of a hormonal system in plants, followed by the evidence of its role in growth and tropisms (Kholodnyi 1928; Went 1928) and the theory of plant development (Chailakhyan 1937), it became clear that phytohormones play an important role both in vegetative growth and in reproductive development. The role of phytohormones in these processes is related to differences in their concentrations in plants of different photoperiodic groups, as well as to the action of environmental factors. For example, Lang (1957, 1965A,B) showed that the treatment of some long-day species (carrot, henbane) with gibberellic acid induces the formation of flowers, despite short-day growth conditions. Thus, the mode of action of the photoperiod on sex expression is related to the functional activity of gibberellins. This conclusion is confirmed in other experiments, in which long days and high gibberellin content brought about the formation of male flowers in cucumber plants, while short days and low levels of gibberellins led to female flowers (Atsmon et al. 1968; Atsmon 1968; Amitawa and Satoru 1970). The general conclusion reached by these and other authors (Wittwer and Bukovac 1958; Bonnet-Masimbert and Nitsch 1969; Splittstoesser 1970; Kaushik and Bisaria 1974) is that high levels of gibberellins and long days determine male sex expression, while high levels of auxins and short days determine female sex expression. In other words, under given photoperiodic conditions, sexual differentiation is regulated by two groups of phytohormones – gibberellins and auxins.[1]

However, Pharis and co-workers (Pharis and Owens 1966; Owens and Pharis 1967, 1971; Pharis and Morf 1968, 1970; Pharis et al. 1970) demonstrated that sex expression in a number of conifers (arborvitae, Arizona cypress, among others) is regulated by day length and the levels of gibberellins only. Specifically, it was established that gibberellin treatments can result in the formation of male

---

[1]Other interpretations are possible. When Jack-in-the-pulpit (*Arisaema triphyllum*) is grown in high light and good soil, mainly female plants are produced; in shade or poor soil, mainly male plants are seen (Lovett-Doust and Cavers: Ecology 63:797–801, 1982). The interpretation is that factors that favor growth and biomass formation lead to femaleness. However, the feminizing effect of darkness makes this unlikely [Ed.].

as well as female strobili. The induction of male primordia necessitated long days and a relatively low concentration of gibberellin (Pharis et al. 1970). On the other hand, the spraying of seedlings of giant arborvitae with a concentrated solution of gibberellin and subsequent growth in 8-hour days in the first year of growth led to the formation of female strobili (Pharis and Morf 1970). Furthermore, Bonnet-Mosimbert (1970, 1971), Dunberg (1972, 1973, 1974), and Pharis et al. (1975) have adduced evidence that the type of sex expression is dependent on the content of specific gibberellins in the tissues of the plant, the relative levels of the polar (A_3, A_8) and nonpolar forms (A_9, A_4, A_7) being especially important. Ivonis et al. (1979) discovered that in the spring, the needles of a clone of Norway spruce that bears a characteristically high number of microstrobili (male cones producing pollen) contain considerably more natural gibberellins than those of another clone of the same species that carries less microstrobili.[1]

Our studies (Chapters 5 and 6) indicate that, in some cases, environmental factors exert their influence on sex expression by altering the size of the organ in which a particular phytohormone is synthesized. Thus, growth in short days stimulates the development of the root system, which results in increased cytokinin levels and thus in female sex expression. On the other hand, growth in long days leads to more developed aerial parts resulting in an increase in the level of gibberellins, and in enhanced male sex expression.

The stimulatory effect of darkness on female sex expression is probably due to the fact that, in the dark, there is a decrease in the content of gibberellins, paralleled by an increase in the levels of cytokinins and in certain inhibitors of a terpenoid nature. Although not yet a general rule, this causal relationship has been demonstrated in a number of experiments in which different species were transferred from long-days to short-days (Wareing and Seth 1967; Wareing et al. 1967; Mizrahi and Richmond 1972; Lozhnikova and Chailakhyan 1975).

The formation of female sexual characteristics as a result of ethylene treatment is also connected to changes in the metabolism and hormonal balance in different plant tissues. Thus, Chrominski and Kopcewicz (1972) conducted an experiment in which Ethrel (220 mg/L) applied to pumpkin plants of the male type led to a dramatic shift towards femaleness (up to 100%). These authors showed that the treatment triggers a sharp decrease in the level of gibberellins (and a slight decrease in the level of auxins). A decrease in the levels of gibberellins and auxins resulting from Ethrel treatment was also observed in an experiment with cucumber (Rudich et al. 1972A). An additional finding in that experiment was that Ethrel triggers an increase in the level of abscisic acid.

Of course, any physiologically active substance acts in conjunction with other regulators. Thus, ethylene is considered to exert its influence on the development and growth of plants by interacting with different phytohormones (Kovalchuk

---

[1]For a recent review on the roles of gibberellins, especially in conifers, see Pharis R.P., and King R.W.: Gibberellins and reproductive development in seed plants. Ann. Rev. Plant Physiol. 36:517–568, 1985 [Ed.].

1977).[1] The treatment of plants with either auxin (Burg 1962) or cytokinins (Fuchs and Lieberman 1968) leads to an increase in the ethylene secretion of the treated plants. No such effect has been observed with either gibberellin (Lieberman and Kunishi 1972) or abscisic acid (Addicott et al. 1969). Ethylene and GA are, in some respects, antagonistic in the effects that they have on plants (Fuchs and Lieberman 1968). The vast majority of experiments indicate that there is a definite trend towards sex transformation that is triggered by different phytohormones. Both in dioecious and in monoecious plants carrying unisexual flowers, gibberellins stimulate male sex expression and cytokinins stimulate female sex expression.

Sex expression in plants is closely linked to environmental conditions, including day length, intensity and quality of light, temperature, humidity, mineral nutrition, and atmospheric gas composition. All these factors can substantially modify certain physiological and biochemical processes, namely the redox system and the synthesis of hormone-like substances. There is good reason to believe that sex transformation and the expression of sexuality involve "secondary" phytohormones and other active substances that independently alter the balance of endogenous "primary" phytohormones, namely gibberellins and cytokinins, which are the most active regulators of sex expression.

A comparative analysis of the evidence for an influence on sex expression by ecological factors on one hand, and by hormonal factors on the other hand, can be most useful in constructing an overall picture of the regulation of sex expression in plants. Figure 7.1 shows how ecological factors can affect the endogenous hormonal system, which in turn interacts with the genetic material, and how the whole system determines the type of sex expression.

The four main environmental factors that influence sexual differentiation in plants are: day length, temperature, mineral nutrition, and humidity (Molliard 1898; Tournois 1911; Schaffner 1919; McPhee 1924; Grishko 1935; Minina 1952; Thompson 1955; Heslop-Harrison 1957; Davidyan 1963; Djaparidze 1965; Maurinya and Berzinya-Berzite 1974; Khrianin and Chailakhyan 1977; Heide 1978). As demonstrated in our own studies, the main phytohormones determining the expression of male and female sexual characteristics are cytokinins that are synthesized in the roots, and gibberellins, produced by the leaves. Environmental factors that favor the synthesis of cytokinins thus lead to female sex expression, while those factors that favor the synthesis of gibberellins stimulate male sex expression.

Short day-length, short wavelength light, the presence of CO, high nitrogen content, high humidity, and relatively low temperatures are all factors that have been found to stimulate the growth of the root system and therefore lead to female sex expression (Dobrunov 1935; Makarevich 1935; Levchenko 1937;

[1]However, in some of the ethylene-sensitive responses (e.g., fruit maturation, petiole epinasty, or the lateral enlargement of longitudinally growing organs), there is as yet no clear evidence that this is the case [Ed.].

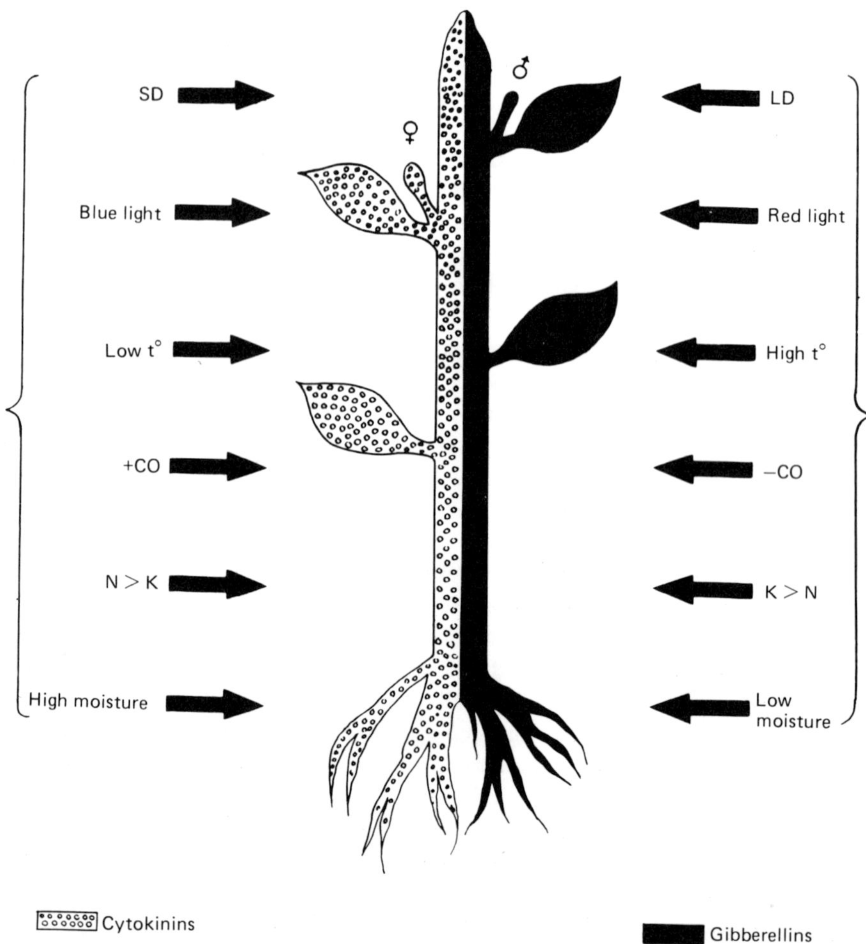

FIGURE 7.1. Roles played by environmental factors and by different phytohormones in sex expression in plants. Left = cytokinins; right = gibberellins.

Kroker 1950; Minina 1952; Ustinova 1956; Heslop-Harrison 1957; Shain 1963; Lebedeva and Yurina 1970; Butnitskyi and Molotkovskyi 1968; Abdel-Gaward and Ketallapper 1969; Molotovskyi 1976). The most specific experiments are probably those conducted by Hewett and Wareing (1973) and Sotta (1978). In these experiments, short day-length was positively identified as a factor that specifically induces the formation and metabolism of a cytokinin (zeatin) in the root system of plants. The general belief, however, is that when such a strong root system produces a large amount of cytokinins they are subsequently transported to the stem apices, where they directly affect the regulatory system in a manner that results in female sex expression.

The preferential growth of the aerial part of the plant, and of the leaves in particular, leads to male sex expression. In this case, the favorable external factors are: long day-length, long wavelength light, high potassium content, high temperature, and relatively low humidity (Kuznetsov et al. 1980). Long days, red, and far-red light stimulate the formation and activity of gibberellins present in the leaves (Lozhnikova 1965; Neskovic and Konjevic 1974). A well-developed leaf system produces large quantities of gibberellins that are transported to the stem apices where they determine male sex expression.

The model built on the experimental evidence gathered in this book leads to a concept of regulation of sex expression in plants that interconnects environmental, hormonal, and genetic factors. On one hand, environmental (or ecological) factors induce changes in the contents of those phytohormones that regulate sex expression; on the other hand, the phytohormones modify the influence of genetic factors. In other words, sex expression in plants is determined by genetic factors that are modulated by the hormonal system that in turn responds to the environment.

8
Practical Uses of Phytohormones in Directing Plant Sex Expression and in Increasing Plant Productivity

The interest in the hormonal regulation of sex expression in dioecious and monoecious plants has led to a number of experiments in growth chambers and in the field. Some of the experiments were even conducted on the commercial scale. This chapter deals with the possible agricultural uses of gibberellins, cytokinins, and Ethrel.

GA can be used in treatment of hemp plants designed for the production of fiber. Indeed, two successive treatments with this phytohormone (the first at the stage of 3 to 4 leaf-pairs, and the second 15 days later using a concentration of 25 mg/L and 400 to 800 L/hectare) result in rapid growth of essentially monoecious plants. The simultaneous maturing of all the plants and their uniformity in height and in the quality of bast fiber represent many advantages. The rapid growth is of paramount importance in temperate climates, such as in temperate regions of the USSR, because the GA-treated hemp can be harvested 2 to 3 weeks before wheat, thus allowing a more efficient use of agricultural machinery.

These beneficial influences of gibberellin on hemp bast production were demonstrated some time ago by a number of researchers (Chailakhyan et al. 1960; Zakordonets 1961; Atal 1959; Zhukov et al. 1963; Zhukov and Sazhko 1963; Davidyan 1967). Unfortunately, the experiments were all conducted on a small scale, i.e., in the growth chamber or on small lots (1.5 to 18 m^2), without detailed studies of the tissues of the stem, nor of the quality of the straw and fiber.

However, a number of experiments, spanning several years, have now been conducted both at semi-industrial[1] and industrial[2] levels. These experiments were aimed at determining the effects of gibberellin on the physiological processes and productivity of hemp, cultivars SOU and US-6 (Khrianin 1966b, 1967). The lot sizes used ranged from 25 to 60,000 m^2 and clearly demonstrated that GA has a desirable influence on the structural elements of the stems (Khrianin 1966a). As a result of GA treatment, the diameter of the bast fibers was doubled, the envelope of bast fibers was thicker by 33%, and the average length

[1]Location: Sovkhoz and Technicum of Penza, USSR.

[2]Location: Lenin Kolkhoz of the Luninskaya Region, District of Penza, USSR.

TABLE 8.1. Effects of gibberellin treatment on growth and productivity of hemp (field studies).[1]

Treatment	Average length of stem (cm)	Average diameter of stem (cm)	Yield of straw (ce/ha)	Yield of fiber (%)	Yield of fiber (ce/ha)	Strength of fiber (kg/cm)	Fiber type (no.)
cv. US-6							
Control	140.0	5.4	38.3	21.3	7.4	30.2	0.7
GA-treated	225.0	6.8	50.2	27.8	11.3	36.5	0.9
cv. SOU							
Control	104.0	5.1	39.0	20.6	6.49	31.6	0.7
GA-treated	248.0	6.5	62.4	21.2	10.69	36.9	0.9

[1]Yields in centners/hectare.

of the fiber was increased by 37%. The resistance of the bast fibers was increased by 12% and the overall yield of fiber by 35%. The productivity was increased by 24% in terms of quantity of straw, and by 35% in terms of quantity of fiber. Finally, the technical qualities of the fiber were also improved (Khrianin 1966b, 1971).

Treating hemp plants with gibberellin does not perturb the mechanisms of vegetative growth *per se*. Therefore, lodging of the crop is not a problem. In fact, the increase in the mass of the hemp plant is due to a balanced increase of photosynthetic and respiratory activities (Khrianin 1964, 1965, 1969) in parallel with a more efficient uptake of nutrients from the soil (Yakushkina and Khrianin 1967). Table 8.1 presents the main experimental results of commercial interest. Figure 8.1 illustrates very well the differences in growth obtained in our field studies. As could be expected, the drastic overall change in the type of flower clusters eventually results in a very poor production of seed. Indeed, the experimentally treated plants yielded 0.5 metric centners[1] of seed per hectare, while the yield was 4.7 centners/hectare in the control plants. Because the material mainly required is fiber, this GA-induced loss (4 to 5 centners/hectare) is not a major disadvantage. The treatment has one final beneficial effect, namely it gives an increase in the productivity of local varieties of hemp, that under natural conditions are less efficient than southern varieties.

Gibberellin is used in the treatment of other plant species as well. In grapes, for instance, GA treatment of plants under conditions that are unfavorable for pollination and fertilization results in parthenocarpy (seedless fruit) in varieties that normally bear the female type of flower. The benefit is that there is an increase in the number of individual berries in the cluster, i.e., an increase in productivity (Manankov 1960, 1963; Tkachenko 1963; Mekhti-Zade 1963; Plakida and Gabovich 1964; Chailakhyan and Sarkisova 1980). In many instances, the GA treatment can replace the very laborious and costly artificial pollination. The

[1]One centner = 100 kg [Ed.].

FIGURE 8.1. An industrial study of the effect of GA. Left = control plants; right = plants treated with GA.

best results are obtained with a single treatment (spraying) during the peak time of flowering of the grapes. In the majority of cases, the concentration of GA of 25 to 50 mg/L of solution is very effective (Chailakhyan 1967; Merzhanian 1967). The benefits of GA treatment were clearly demonstrated in large-scale experiments in which the phytohormone consistently improved the productivity of grapes, of the cultivar Tchaush (Manankov 1960, 1963).[1]

GA treatment of tomatoes is also quite useful. Spraying of female, self-pollinating lines leads to the production of hybrid seeds. GA induces the differentiation of anthers and leads to the formation of pollen in astaminate, mutant types of tomato (Phatak et al. 1966).

In some cases, gibberellins restore the fertility of barley pollen, which leads to the fertilization of flowers with male sterility. On the other hand, the massive spraying of young corn plants (before flowering) results in the sterility of pollen (Muromtsev and Agnistikova 1971; Khrianin 1978). Moreover, gibberellin also induces the formation in corn of hermaphroditic flowers (Sladky 1969, 1971; Krishnamoorthy and Talukdar 1976). The GA treatment stimulates the growth of pollen and the elongation of pollen tubes (Khrianin 1969).

---

[1]Formation of parthenocarpic grapes and their increase in size after GA treatment were probably first reported by Weaver at Davis, California in 1958 (see Plant Growth Substances in Agriculture, by R.J. Weaver, San Francisco, CA, W.H. Freeman and Co., 1972) [Ed.].

Gibberellin (75 to 250 mg/L) can be used to induce the formation of male cones in many conifers (Pharis and Owens 1966, 1967; Pharis and Morf 1970; Pharis et al. 1970, 1974, 1975). If GA is used in very high concentrations (2500 mg/L), however, female cones appear in the same conifers that can be induced to form male ones. It is interesting that gibberellin A_3, in combination with A_4, A_7, and A_{13}, induces the formation of cones in giant *Sequoia* at an age as young as 8 months (Pharis and Morf 1972).

Gibberellins can be of use in the planting of greenery in cities. It is well known that female poplar trees yield an abundant quantity of seeds that litter streets, parks, and open places with down-like material. This problem is conventionally dealt with by cutting the crowns of poplar trees in winter and early spring. An elegant alternative to this method would be a treatment of female trees with gibberellin to produce predominantly male flowers, as these do not form down-like structures.[1]

The stimulation of female sex expression as a result of cytokinin treatment is useful in the treatment of agriculturally important plants, such as those in the *Cucurbitaceae*. For example, the introduction of an analogue of cytokinin, 6-BAP, into the root system of young cucumber seedlings results in a threefold increase in the number of pistillate flowers (concentration of the analogue: 10 to 15 mg/L; time of treatment: 24 to 28 hours). It should be noted that 6-BAP causes the formation of female flowers in the very first internodes that normally carry only male flowers.[2] Natural and synthetic cytokinins also have found applications in the production of grapes. Male clones or shoots are kept in a solution of 6-BAP (1 mg/L), for 3 weeks or the shoots are simply treated with a solution of zeatin (1 to 10 mg/L). The result is the formation of both functionally male and bisexual flowers (Negi and Olmo 1966, 1972; Moore 1970; Hashizume and Iizuka 1971).

Ethylene, that can be produced from several compounds, is another regulator of sexual differentiation in plants (Rakitin and Rakitin 1977). Its artificial precursor, Ethrel (2-chloroethylphosphonic acid)[3] is becoming more and more widely used. Ethrel stimulates the formation of female flowers, while inhibiting male sex expression in many *Cucurbitaceae*, such as cucumber, melon, and pumpkin (Iwahori et al. 1969, 1970; Tarakanov and Agapova 1973, among others). Moreover, Ethrel can be used in the production of hybrid seeds in cucumber (Agapova 1975; Tronichkova 1978) and in melon (Sankin et al. 1978). It is also efficient in inducing male sterility in cereal crops (Hughes et al. 1974; Fedin and Gyska 1975). The treatment of *Cucurbitaceae* usually requires a concentration of

---

[1]The same application could apply to gingko trees, the plentiful fruits of which produce a disagreeable odor when rotting on the ground. Gingko is often planted as a street tree in the United States [Ed.].

[2]Some genetic lines of cucumber are completely female and, in these, GA is applied to produce the male flowers needed for seed production (Acta Horticulturae 31:81–82, 1973) [Ed.].

[3]Also called Ethephon. The natural precursor, 1-aminocyclopropane-1-carboxylic acid (ACC), is also coming into use now [Ed.].

Ethrel of 0.1–0.6 mg/L, while wheat and barley plants are sprayed with a 300 to 600-mg/L solution. An increase in the commercial production of plants of the gourd family grown in greenhouses is routinely achieved by Ethrel treatments.[1]

A recent development in the production of cucumbers in winter greenhouses is the treatment of plants with "dihydrel." The seedlings (stage of 2 to 3 leaf-pairs) are sprayed with a 0.01% solution of "dihydrel" (5 mL per plant). This treatment stimulates the development of pistillate flowers with an eventual increase in productivity of 20% to 29% (Budykina et al. 1980).

It should be noted, finally, that gibberellins, cytokinins, and ethylene are not only used in regulating sex expression. These substances have found many other applications in different areas of agriculture.

[1]Auxin often has the same effect [Ed.].

9 Conclusion

For a long time, the study of sexuality in plants consisted mainly of measuring the influence of external factors on sexual dimorphism and in determining the physiological and biochemical differences between male and female plants. The recent expansion of this area of plant biology is due to the discovery of phytohormones and inhibitors, followed by experiments demonstrating that these new substances participate not only in growth and development, but also in sexual differentiation. As a result of the many different experimental approaches, it has been possible to bring to light some of the mechanisms of hormonal regulation of sex expression in plants, and to determine the causes of sexual differentiation in dioecious plants and in monoecious plants that bear unisexual flowers.

The general conclusion from the early experiments, carried out by many investigators, was that there is a clear tendency towards sex transformation when experimental plants are sprayed with various solutions of phytohormones. In this older work, gibberellins shifted the sex ratio in favor of male individuals (or male flowers), while auxins and, to some extent, cytokinins shifted the sex ratio towards femaleness. However, because of the method of treatment (i.e., spraying) the results lacked consistency and were difficult to reproduce. The experimental plants were sprayed at different ages and at different developmental stages, and the endogenous levels of phytohormones were never taken into consideration. Thus, it was necessary to determine the age at which the hemp and spinach stem apices are most sensitive to the action of the phytohormones. This amounted to determining the critical period during which flower primordia initiate differentiation.

The anatomical studies turned out to be quite successful in settling this first stage of the problem. The sexual differentiation of stem apices occurs very early in the development of both plants: at the stage of three leaf-pairs in hemp, and at the stage of three leaves in spinach. It is at that time that the stem apices are most sensitive to environmental factors and to phytohormones. Thus, it became clear that in order for the phytohormones to be effective, they had to be introduced at the earliest possible stage of development.

This conclusion was initially supported by experiments in which the seeds themselves were treated before sowing, and later by the culture of isolated embryos in White's medium with and without phytohormones. The results obtained in culture were particularly conclusive: the addition of gibberellin led to 100% male sex expression in hemp, and the addition of 6-benzylaminopurine (BAP) resulted in 97% of the embryos developing into female plants.

The next step was to determine the roles of the individual organs in sexual differentiation. It was already suspected that male sex expression was connected with the development and activity of the aerial parts, while female sex expression involved the root system (Minina 1952; Molotkovskyi 1960). To obtain direct evidence, we developed an integral model of sex expression in plants which was subsequently put to experimental trial. It was discovered that both the roots and the leaves play important roles in this phenomenon; the roots are responsible for female sex expression, and the leaves for male sex expression. The next question addressed the nature of the compounds necessary for male and female sex expression. The effects observed when 6-BAP was introduced into plants with their roots removed indicated that the action of the roots in promoting female sex expression and the synthesis of cytokinins in the root system were closely related. The experimental introduction of GA_3 into plants with their leaves removed demonstrated that the mode of action of leaves in promoting male sex expression involved the synthesis of gibberellins in the leaf tissues.

These findings were in complete agreement with analytical studies of the variations in the endogenous levels of the two phytohormones. Female plants have significantly higher levels of cytokinins than male plants, whereas male plants contain much more gibberellin than do females. Moreover, the well-established connection between the metabolism of roots and leaves (Sabinin 1949; Kursanov 1960, 1976) was clearly observed in the studies of the hormonal balance in intact and mutilated plants. Specifically, it was established that the levels of endogenous cytokinins and gibberellins are not independent variables. High cytokinin activity is paralleled by low gibberellin activity, and vice-versa. This finding supports the notion that the roots, being producers of cytokinins, play an important role in leaf metabolism, and that the leaves, in their turn, being producers of gibberellins, influence the metabolism of the roots (Kulaeva 1973; Chung and Chung 1978). Thus the physiological roles played by the phytohormones in plant development appear to be crucial (Chailakhyan 1975; Krekule and Seidlova 1977; Sotta 1978).

The interaction between the individual organs that synthesize different phytohormones is very important in determining sex expression, for gibberellins and cytokinins actively migrate through the vascular system that connects the leaves and the roots to the shoot (Carr and Burrows 1966; Torrey et al. 1977). According to Miginiac (1978), the roots play an important role in the correlations that are regulated by the functions of the meristem. A similar conclusion can be reached about the leaves. Cytokinins are produced predominantly[1] in the roots and are subsequently transported to the stem apices. In these, the cytokinins activate the

[1]However, there is evidence that smaller amounts are formed also in buds [Ed.].

functioning of the meristem and apparently induce the regulatory system that determines female sex expression. The situation is similar in the case of gibberellins that are synthesized in the leaves. Gibberellins are transported from the leaves to the apical buds where they are activated. Apparently, gibberellins mainly act in conjunction with other phytohormones (Khrianin and Chailakhyan 1979).

The modes of action of these two phytohormones were studied in experiments in which the introduction of the phytohormones was followed by treatment with inhibitors of nucleic acid and protein metabolism. Mitomycin and puromycin interfere with female sex expression caused by 6-BAP, which implies that cytokinins are active at the level of replication and translation. Actinomycin D, however, counteracts the effect of gibberellin in stimulating male sex expression, which implies that gibberellins act at the level of transcription. In fact, these results are the first tangible indication that the mode of action of the phytohormones in sex expression involves changes in the activity of the genetic apparatus of the cell.[1]

The tissues of female plants contain specific cytokinins and a specific protein that are absent from male plant tissues. This finding is owed to chromatographic studies and immunochemical analysis. Thus, it is possible that there is a cell-based interaction between specific cytokinins and proteins, perhaps a formation of a hormone-receptor complex,[2] which regulates the expression of the genes responsible for female sex expression. Following this logic, the gibberellins synthesized in the leaves would be transported into the cells of the apical meristem where they would bind to specific receptors and possibly to other proteins that can activate the complete developmental program of male sex expression. Perhaps the hormone-receptor complexes act on gene-modifiers that regulate the level of male or female sex expression. Interestingly, the existence of gene modifiers has been demonstrated in many plants, such as corn, hemp, and grape (Jones 1939; Köhler 1964; Negi and Olmo 1972).

Naturally, sex expression is not regulated by cytokinins and gibberellins alone. Sexual differentiation should be seen as a chain of interacting events that are triggered by one or many internal and external factors. This is clearly indicated by our studies of the biological activity of phytohormones and of the effects of individual and combined treatments with phytohormones and inhibitors. However, there appear to be two main factors (the most direct causes) that trigger the differentiation of sexes. These two factors are, of course, cytokinins and

---

[1]In the case of cell enlargement caused by auxin, the evidence was adduced much earlier (see Noodén L.D., Thimann K.V.: Proc. Natl. Acad. Sci. USA 50:194–200, 1963; Noodén L.D., Thimann K.V.: Plant Physiol. 40:193–201, 1965; Key J.L.: Plant Physiol. 39:365–370, 1964) [Ed.].

[2]A number of recent studies have been devoted to the elucidation of cytokinin-receptor complexes (for example, see Polya G.M., Bowman J.A.: Plant Physiol. 64:387–392, 1979; Erion J.L., Fox J.E.: Plant Physiol. 67:156–162, 1981; Brinegar A.C., Fox J.E.: Biol. Plantarum 27:100–104, 1985) [Ed.].

gibberellins. These most effective, primary phytohormones act in cooperation with other, secondary but nonetheless important phytohormones (Khrianin and Chailakhyan 1979). The most important secondary phytohormones are auxins and ethylene . Heslop-Harrison (1956, 1957, 1963) attached great importance to the plant sex transformation that results from the influence of auxins. He proposed that high levels of auxins stimulate the formation of female flowers, while low levels of auxins induce the formation of male flowers. Auxin had a stimulatory effect on the formation of female flowers in our experiments as well, although the effect of cytokinins was stronger. Ethylene treatment also results in a significant shift towards femaleness. This effect of ethylene on sex expression can be explained by the close connection that this compound has both with auxins (Burg 1962; Torrey et al. 1977) and cytokinins (Fuchs and Lieberman 1968). Thus it appears likely that ethylene is affecting sex expression by somehow modifying the levels of cytokinins in the plant.

The mode of action of ABA is still very poorly understood because of the incomplete data and ambiguous interpretations. Nevertheless, the majority of studies (Abdel-Gaward and Ketallaper 1969; Rudich et al. 1972a; Rudich and Halevy 1974), including our own experiments, indicate that ABA stimulates female sex expression in dioecious plants and in monoecious plants that carry unisexual flowers. Since ABA certainly counteracts the stimulatory effect of exogenous gibberellins on male sex expression (Mohan Ram and Jaiswal 1972), it is highly probable that it has the same effect on natural, endogenous gibberellins. Its action may perhaps be linked to changes in the level of proline, which may play an important role in the shift from vegetative to reproductive growth and may act later also in the fertilization processes (Britikov 1975; Martin 1977).

A connection between the specific process of sexual differentiation and the general one of growth and development has been emphasized by a number of authors (Sidorskyi 1972; Lvova 1971; Minina and Larionova 1979). Some experiments seem to indicate that rapid growth favors female sex formation, while slow growth results in male sex expression (Pravdin 1950; Minina 1954; Tompsett 1978). Minina (1949) has directly demonstrated that an increase in the growth rate of experimental plants has a stimulatory effect on the formation of female flowers. In fact, a direct correlation between the high level of auxins, the growth rate, and female sex expression was revealed in several experiments (Zimmerman 1936[1]; Minina 1960; Heslop-Harrison 1963; Minina and Larionova 1979). Unfortunately, no connection could be established in a number of other experiments (Avery et al. 1937; Culafić and Nesković 1974; among others). Moreover, there are contradictory results indicating stronger female sex expression associated with slower growth (Uglov 1959; Sidorskyi and Sidorskaya 1970).[2]

---

[1]Perhaps even more clearly in the male, female, and parthenocarpic flowers of squash (Nitsch et al.: Am. J. Bot. 39:32–43, 1952) [Ed.].

[2]The work on hemp described in the preceding chapter also shows that the increased growth due to GA treatment is associated with predominance of maleness [Ed.].

In our experiments on the effects of IAA and other auxins, the auxin analogue (at the concentration used) practically never stimulated the growth in height of the treated plants, despite the fact that female sex expression was favored. However, the male sex expression due to gibberellin treatment was always accompanied by an increase in the height (stem elongation), while the induction of female sex expression by cytokinins caused partial inhibition of the processes of growth and development. Still, it is difficult to connect the specific action of phytohormones on sex expression with the general processes of growth.

The general conclusion from the experiments presented in this book is that sex determination is indeed a function of the genetic apparatus, while sex expression is strongly affected by the hormonal balance that exists under natural and experimental conditions.

Under natural conditions and in the absence of extreme situations the activities of the individual components of the hormonal system, notably cytokinins and gibberellins, are tuned in a manner that ensures approximately equal proportions of males and females in a population of a dioecious species. Conditions that cause high levels of synthesis of gibberellins in the leaves eventually lead to an increase in the number of male plants. On the other hand, factors that cause the root system to produce more cytokinins lead to female sex expression. According to the proposed unifying concept, there are two main stages in the chain of events that constitute sex expression. In the first stage, the receptor organs of the plant respond to ecological factors by altering the hormonal system. In the second stage, the endogenous hormonal factors in their turn influence the genetic apparatus.

At present, we only have a rough outline of the pattern of interactions thus involved in the regulation of sex expression in plants. There is no doubt that there are exciting prospects for the future. Understanding of the mechanisms of sex determination and sex expression will be of great importance in learning how to direct sexual differentiation and how to increase the productivity of useful plants.

The findings made in the study of sex expression in dioecious plants and in monoecious plants bearing unisexual flowers may perhaps eventually be applied to monoecious plants with bisexual (hermaphroditic) flowers, where the processes of male and female sexual differentiation are intimately connected. The study of the interactions between the hormonal system and the genetic apparatus will no doubt remain one of the most exciting areas of modern plant biology.

References*

Abdel-Gaward H.A., Ketellapper H.J.: Regulation of growth, flowering and senescence of spinach plants. II. Effects of 2-chlorethylphosphonic acid (Ethrel) and abscisic acid. Plant Physiol. 1969, 44: 15.

Abelev G.I., Avenirova E.A.: Isolation of precipitating antibodies against specific antigens of mouse liver and hematome. Vopr. Onkol. 1960, 6: 57–62. (R)

Addicott F.T., Lyon J.L.: Physiology of abscisic acid and related substances. Ann. Rev. Plant Physiol. 1969, 20: 139–164.

Addicott F.T., Carns H.R., Lyon J.L., Smith O.E., McMeans J.L.: On the physiology of abscisins. Régulateurs naturels de la croissance végétale. Colloq. Internat. CNRS, 1969, 123: 687.

Agapova S.A.: Characteristics of seed production in heterotic hybrids of cucumber using physiologically active substances. Dokl. TSKhA. 1975, 211: 78–83. (R)

Akkos E.: Zytogenetische Beitrage zur Geschlechtserrebung beim Spinat (*Spinacia oleracia*) im Hinblick auf die Zuchtung triploider Hybridsorten. Z. Pflanzenzucht, 1965, 53: S. 226–246.

Aleshin E.P., Aleshin N.E., Avakyan E.R., et al.: On the mode of action of gibberellin on rice. Fiziol. rastenyi, 1978, 25: 262–267. (R)

Allen C.: Sex inheritance and sex determination. Amer. Nat. 1932, 64: N 703, 97–107.

Amitawa B, Satoru T.: Effect of gibberellin upon sex expression and internode length in gynoecius and monoecious cucumber. J. Jap. Soc. Hort. Sci., 1970, 39: N 3, 224–231.

Andreenko S.A., Kuperman F.M.: Physiology of Corn. M.: Izd-vo MGU. 1959, p. 290.

Anisimov V.V.: Fotoperiodizm konopli (*Cannabis sativa*): Avtoref. diss. kand. biol. nauk. L.: AFN 1966, 22. (R)

Anisimov V.V.: The reaction of hemp plants to photoperiodic changes. Sb. trudov aspirantov i molodykh nauchnykh sotrudnikov. L.: VIR, 1967, 8: 207–212. (R)

Arinshtein A.I., Loseva Z.G.: Selection of monoecious hemp and the determination of conditions favoring the expression of monoecism. Tr. po prikl. botanike, genetike, selektsii, 1958, 31: 201–210. (R)

Aristotle: Collected Works (in four volumes). M.: Mysl, 1975, 1: 550.

Astaurov B.L.: Experimental androgenesis and hypogenesis in mulberry silkworm. Biol. Zhurn. 1937, 6: 3–50. (R)

*Papers in Russian are marked (R) (Ed.).

Astaurov B.L.: Artificial Parthenogenesis in the Mulberry Silkworm. M.; L.: Izd-vo AN SSSR, 1940, p. 238. (R)

Astaurov B.L.: The question of sex regulation. In: Science and Mankind. M.: Znanie, 1963, 2: 344–367. (R)

Atal C.K.: Sex reversal in hemp by application of gibberellin. Current Sci., 1959, 28: N 10, 408–409.

Atkinson G.F.: Experiments on the morphology of Arisaema. Bot. Gaz., 1898, 25: S. 114.

Atsmon D.: The interaction of genetic, environmental and hormonal factors in stem elongation and floral development of cucumber plants. Ann. Bot., 1968, 32: N 128, 877–882.

Atsmon D., Galun E.: Physiology of sex in *Cucumis sativus L.* Leaf age patterns and sexual differentiation of floral buds. Ann. Bot. 1962, 26: 102–105.

Atsmon D., Lang A., Light E.N.: Contents and recovery of gibberellins in monoecious and gynoecious cucumber plants. Plant Physiol., 1968, 43: N 5, 806–810.

Avery G.S., Burkholder R., Creighton H.B.: Production and distribution of growth hormone in shoots of Aesculus and Malus and its probable role stimulating cambiae activity. Am. J. Bot. 1937, 24: N 1, 54–58.

Badr S.A., Hartmann H.I.: Effect of fluctuating vs. constant temperature on flower induction and sex expression in the olive (*Olea europea*). Physiol. Plant., 1971, 24: 40–45.

Barnes M.F., Light E.N., Lang A.: The action of plant growth retardants on terpenoid biosynthesis. Inhibition of gibberellic acid production on *Fusarium moniliforme* by CCC and Amo-1618, action of these retardants on sterol biosynthesis. Planta, 1969, 88: N 2, S. 172–182.

Bazin M., Chabin A., Durand R.: Comparison between four isoaccepting transfer ribonucleic acids and corresponding synthesis in male and female flowers of the dioecious species *Mercurialis annua L.* Develop. Biol., 1975, 44: N 2, 288–297.

Belar K.: Der Formwechsel der Protistenkern. Eine vergleichendmorphologische Studie. Jena, 1926, S. 106.

Belyaev D.K., Klochkov D.V., Zhelezova A.I.: The influence of light conditions on the reproductive functions and fertility in mink (*Mustela vison Schr.*). Bul. MOIP, Otd. biol., 1963, 68: 107–125. (R)

Bennet M.D., Hughes W.G.: Additional mitosis in wheat pollen induced by ethrel. Nature (London), 1972, 240: 566–568.

Berg L.S.: Patterns in the formation of organic forms. Tr. po prikl. botanike, genetike, selektsii, 1925, 14: 19–68. (R)

Bessey E.A.: Effect of the age of pollen upon the sex of hemp. Am. J. Bot. 1918, N 5, 234–238.

Bessey E.A.: Sex problem in hemp. Quart. J. Biol., 1933, 3: 185–190.

Bhandari M.C., Sen D.N.: Effect of certain growth regulators in the sex expression of *Citrullus lanatus (Thunb.)* Munsf. Biochem. und Physiol. Pflanz., 1973, 164: N 4, S. 450–453.

Biddington N.L., Thomas T.H.: A modified Amaranthus Betacyanin bioassay for the rapid determination of cytokinins in plant extracts. Planta, 1973, 111: N 2, S. 185–185.

Bigot M.C.: Action d'adénines substituées sur la synthèse des bétacyanines dans la plantule d'*Amaranthus caudatas L.* Possibilité d'un test biologique de dosage des cytokinines. C. r. Acad. sci. (Paris), 1968, D 266: 349–351.

Bisaria A.K.: The effect of foliar spray of alpha naphthaleneacetic acid on the sex expression in *Monordica charantia L.* Sci. and Cult., 1974, 40: 78–80.

Bobrov E.G., Karl Liney L.: Nauka, 1970, 286. (R)

Bonner J.: Molecular Biology of Development. M.: Mir, 1967, p. 180. (R)

Bonner J.: Plant Biochemistry. M.: Mir, 1968, 624. (R)

Bonnet-Masimbert M.: Artificial induction of male and female flowers on young seedlings of *Cupressus Arizonica (Creene)* and *Chamaecyparis Lawsoniana (Parl.)*. Sexual reproduction of forest trees. Finland, Proc. Meet. Varparants, 1970, S. 189.

Bonnet-Masimbert M.: Induction florale précoce chez *Cupressus arizonica* et *Chamaecyparis Lawsoniana*. Silvae Genet. 1971, 20: N 13, S. 82–90.

Bonnet-Masimbert M., Nitsch J.P.: Induction précoce de la floraison chez une plante ligneuse *Rhus typhina L.* Bull. Soc. Bot. France, 1969, 116: N 9, 403–408.

Borodulina F.Z.: On the influence of mineral nutrition on sex transformation in cucumber. In: Sb. rabot stud. kruzhkov. M.: Izd-vo MGU, 1938, v. 2, p. 72. (R)

Borthwick H.A., Sally N.J.: Photoperiodic responses of hemp. Bot. Gaz., 1954, 116: 14–29.

Bose T.K., Nitsch J.P.: Chemical alteration of sex expression in *Luffa acutangula*. Physiol. Plant., 1970, 23: 1206–1211.

Bosse G.G.: Artificial sex transformation in Eucommia. Sov. subtropiki, 1935, 7: 64–68.

Boyarkin A.N.: A method for quantitative determination of the activity of growth substances. In: Methods for identifying growth regulators and herbicides. M.: Nauka, 1966, pp. 13–15. (R)

Boysen-Jensen P.: Plant Growth Hormones. M.; L.: Biomedgiz, 1938, p. 251.

Brantley B.B., Warren G.F.: Sex expression and growth in muskamelon. 1960, 35: 741–745.

Breslavets L.P.: Cytological studies of bast plants. 1. *Cannabis sativa* (hemp). Tr. In-ta novykh lubyanykh kultur, 1933, 2: 112–116.

Breslavets L.P.: Studies of the development of flowers of hemp plants that have undergone a photoperiod-induced sex change. Bull. MOIP, 1936, 14: 182–189. (R)

Britikov E.A. The Biological Role of Proline. M.: Nauka, 1975. (R)

Brown-Séquard M.: C. r. Soc. Biol. 1889, 41: p. 420.

Budykina N.P., Drozdov S.N., Volkova R.I., et al.: The use of dihydrel in growing cucumber in winter greenhouses. Inform. listok No. 193-80 Karel. Ts.N.T.I. Petrozavodsk, 1980, p. 4. (R)

Burg S.P.: The physiology of ethylene formation. Ann. Rev. Plant Physiol., 1962, 13: 165–302.

Burgeff H, Seybold A.: Zur Frage der biochemischen Untersuchungen der Geschlechter. Ztschr. Bot., 1927, N 19, S. 497–499.

Butenandt A., Jacobi H.: Über die Darstellung eines krystallisierten pflanzenlichen Tokokinins (Thelykinins) und seine Identifizierung mit dem α-Follikelhormon. Ztschr. Physiol. Chem., 1933, 218: S. 104–106.

Butenko R.G.: Plant Tissue Culture and Physiology of Plant Morphogenesis. M.: Nauka, 1964. (R)

Butnitskyi I.M., Molotkovskyi G.Kh.: The effect of the photoperiod on the growth of aerial and subterranean organs in hemp plants of different sexes and the accumulation of protein. Ukr. Botan. Zhurn., 1968, 25: 22–27. (R)

Byers R.E., Baker L.R., Sell H.M., et al.: Ethylene: a natural regulator of sex expression of *Cucumis melo L.* Proc Natl. Acad. Sci. USA, 1972, 69: 717–720.

Camerarius R.J.: De sexu plantarum epistola. Tuebingae, 1694, S. 25.

Camp W.H.: Sex in Arisaema triphyllum. Ohio J. Sci., 1932, 32: 147–149.

Carr D.J., Burrows W.J.: Evidence for the presence in xylem sap of substances with kinetin-like activity. Life Sci., 1966, N 5, 2061–2077.

Carr D.J., Beid D.M.: The physiological significance of the synthesis of hormones in roots and of their export to the shoot system. In: Sixth Intern. Conf. of Plant Growth Substances: Book of Abstracts. Ottawa, 1967, pp. 10–12.

Catarino F.: Some effects of kinetin on sex expression in *Bryophyllum crenatum*. Port acta biol., 1964, A8: 267–284.

Chailakhyan M.Kh.: Hormonal Theory of Plant Development. M.; L.: Izd-vo AN SSSR, 1937, p. 198.

Chailakhyan M.Kh.: The effects of the environment and internal factors in the flowering of plants. Tr. Erevan. un-ta, 1943, 22: 37–70. (R)

Chailakhyan M.Kh.: The development of different species of Orobanche in relation to the development and growth of the host plants. Dokl. Akad. Nauk. SSSR, 1947a, 4: 881–884. (R)

Chailakhyan M.Kh.: The false immunity of plants. Dokl. Akad. Nauk. SSSR, 1947b, 61: 99–101. (R)

Chailakhyan M.Kh.: Susceptibility to flower parasites and sexuality in plants. Dokl. Akad. Nauk. SSSR, 1947c, 61: 321–332. (R)

Chailakhyan M.Kh.: Hormonal factors in plant flowering. Fiziol. rastenyi, 1958a, 5: 541–560. (R)

Chailakhyan M.Kh.: Principal Characteristics of Ontogenesis in Higher Plants. M.: Izd-vo AN SSSR, 1958b. (R)

Chailakhyan M.Kh.: Factors in the generative development of plants. XXV Timiryazevskie chtenia. M.: Nauka, 1964, p. 58. (R)

Chailakhyan M.Kh.: A Manual on the Use of Gibberellins on Vines. M.: Kolos, 1967. (R)

Chailakhyan M.Kh.: The hormonal regulation of flowering in plants of different photoperiodic groups. Fiziol. rastenyi, 1971, 18: 348–357. (R)

Chailakhyan M.Kh.: Integrity and differentiated models of flowering in plants. In: Biology of Plant Development. M.: Nauka, 1975, pp. 24–47. (R)

Chailakhyan M.Kh., Kochankov V.C., Zamota V.P.: The effects of gibberellin on growth and productivity in hemp and tobacco. Fiziol. rastenyi, 1960, 7: 340–343. (R)

Chailakhyan M.Kh., Martinovich L.N., Kochankov V.G.: The chemical regulation of the growth and formation of generative organs in dioecious plants. Dokl. Akad. Nauk. SSSR, 1969, 189: 662–665. (R)

Chailakhyan M.Kh., Egorova T.A., Yanina L.I.: The effects of darkness and of the retardant CCC on growth and flowering in short-day species. Dokl. Akad. Nauk. SSSR, 1970, 51: 244–248. (R)

Chailakhyan M.Kh., Nekrasova T.V.: The antagonism between gibberellin and retardants in their effects on flowering and fruit-bearing in lemon. S.-kh. biologia, 1976, 11: 48–52. (R)

Chailakhyan M.Kh., Kochankov V.G., Muzafarov B.M.: Uninterrupted darkness as an inducer of stem formation and flowering in the rosulate plant Rudbeckia. Dokl. Akad. Nauk. SSSR, 1976, 231: 1497–1500. (R)

Chailakhyan M.Kh., Grigorieva N.N., Lozhnikova V.N.: The effects of leaf extracts from flowering tobacco plants on the flowering of seedlings in *Chenopodium rubrum L.* Dokl. Akad. Nauk. SSSR, 1977, 236: 773–776. (R)

Chailakhyan M.Kh., Kochankov V.G., Muzafarov B.M.: The effects of uninterrupted darkness on stem formation and flowering in the rosulate plant Rudbeckia combined with long day induction. Dokl. Akad. Nauk. SSSR, 1977, 234: 252–255. (R)

Chailakhyan M.Kh., Khrianin V.N.: Hormonal regulation of sex expression in plants. Botan. zhurn., 1980, 65: 153–171. (R)

Chailakhyan M.Kh., Sarkisova M.M.: Growth Regulators in Vines and other Crops. Erevan: Izd-vo AN ArmSSR, 1980, 187. (R)

Chailakhyan M.Kh.: Internal factors of plant flowering. Ann. Rev. Plant Physiol., 1968, 19: 1–36.

Chailakhyan M.Kh.: Hormonal regulation of plant flowering of different photoperiodical groups. In: Proc. Intern. Conf. Canberra. Plant Growth Substances, 1970, B. New York, Springer-Verlag, 1972, S. 745–752.

Champault A.: Etude caryosystématique et écologique de quelques Euphorbiacées herbacées et arbustives africaines. Bull. Soc. Bot. France, 1970, 117: N 4, 137–168.

Champault A.: Effets de quelques régulateurs de la croissance sur des noeuds isoics de *Mercurialis annua L.* (2n-16) cultivés in vitro. Bull. Soc. Bot. France, 1973, 120: N 3/4, 87–99.

Chekmin I.F., Biglov T.T., Akhmetov R.R., et al.: The influence of low temperatures on the RNA of embryos of rye and wheat. In: Biology of Nucleic Acid Metabolism in Plants. M.: Nauka, 1964, 130–135. (R)

Chrelashvili M.N.: The influences of water content and of assimilate accumulation in the leaves on the energy of photosynthesis and respiration. Eksperim. botanika, 1941, 5: 101–137. (R)

Chrelashvili M.N., Djaparidze L.I.: Differences in the respiration of wintering and growing branches in dioecious plants. Soobsh. GSSR, 1950, 11: 297–301. (R)

Choudhri R.S., Krishan R.: Sex differentiation in *Zea mays*. Sci. Cult., 1946, 11: 472–475.

Chrominski A., Kopcewicz J.: Auxins and gibberellins in 2-chloroethylphosphonic acid–induced femaleness of *Cucurbita pepo. L.* Z. Pflanzenphysiol, 1972, 68: S. 184–189.

Chulafich L.N.: The influence of growth regulators on sex expression in spinach. In: Proceedings of the International Conference on Plant Growth Factors. Liblitse, 1978, 60.

Chung Pei Hsia, Chung Hueii Kao: The importance of roots in regulating the senescence of Soybean primary leaves. Physiol. Plant., 1978, 43: 385–389.

Ciesielski Th.: Quomodo fiat ul max proles masculina, mox feminina oriatur apud plantas, animalia et homines. Lwow (Lemberg), 1911, S. 1–15.

Clark R.K., Kenney D.S.: Comparison of staminate flower production on gynoecious strains of cucumbers, *Cucumis sativus L.* by pure gibberellins (A_3, A_4, A_7, A_{13}) and mixtures. J. Am. Soc. Hort. Sci., 1969, 94: 131–132.

Corley R.H.: Sex differentiation in oil palm: Effects of growth regulators. J. Exp. Bot., 1976, 27: N 98, 553–558.

Correns C.: Gregor Mendels Regel über das Verhalten der Nachkommenschaft der Rassenbastarde. Ber. Dtsch. bot. Ges., 1900a, 18: S. 158.

Correns C.: Uber Levkojenbastarde. Zur Kenntnis der Grenzen der Mendelschen Regeln. Bot. Zentr.-Bl., 1900b, 84: S. 97.

Correns C.: Weitere Beiträge zur Kenntnis der dominierenden Merkmale und der Mosaikbildung der Bastarde. Die Merkmalspaare bei Studium der Bastarde. Ber. Dtsch. bot. Ges., 1902a, 21: S. 195.

Correns C.: Über den Modus und den Zeitpunkt der Spaltung der Anlagen bei den Bastarden von Erbsen. Typus. Bot. Ztg 1902b, 60: S. 65.

Correns C.: Die Vererbung der Geschlechtsformen bei den gynodiözischen Pflanzen. Ber. Dtsch. bot. Ges., 1906, 24: S. 459–474.

Correns C.: Die Bestimmung und Vererbung des Geschlechts nach neuen Versuchen mit höheren Pflanzen. B.: Borntraeger, 1907, S. 27.

Correns C.: Über nichtmendelnde vererbung. Ztschr indukt. Abstammungs-und Vererbungslehre, 1928a, Suppl., S. 131–168.
Correns C.: Bestimmung, vererbung und Verteilung des Geschlechtes bei den höheren Pflanzen. Hdb. Vererbungswiss. B.: Borntraeger, 1928b, S. 31.
Cross B.E., Myers P.L.: Effect of plant retardants on the biosynthesis of *Gibberella fujikuroi*. Phytochemistry, 1969, 8: N 1, 79–83.
Crozier A.: The role of root apices in gibberellin metabolism of *Phaseoulus multiflorus* seedlings. In: VII Intern. Conf. on Plant Growth Substances: Abstracts. Canberra, 1970, p. 19.
Cŭlafić Lj., Nesković M.: Indole auxines in spinach plants grown in long and short days. Biol. Plant. Acad. Sci. Bohemosl., 1974, 16: S. 359–365.
Currence T.M.: Nodal sequence of flower type in cucumber. Proc. Amer. Soc. Hort. Sci., 1932, 29: 447–451.
Davey A.J., Gibson C.M.: Note on the distribution of the sexes in *Myrica gale*. New Phytol., 1917, 16: 147–151.
Davidyan G.G.: The growth and development of hemp as a function of light conditions. Agrobiologia, 1963, 6: 922–925. (R)
Davidyan G.G.: The effect of gibberellin on the growth, development, and sex formation in hemp. S-kh. biologia, 1967, 2: 90–94. (R)
Davidyan G.G., Kutuzova S.N.: On organogenesis in hemp (*Cannabis sativa L.*). Bul. VIR, 1970, 15: 69–73. (R)
Davidyan G.G., Rumyantseva L.T.: Ethrel induction of female flowers on male hemp plants. Vestn. s.-kh. nauki, 1974, 4: 49–51. (R)
Dadykin V.P.: The Characteristic Behavior of Plants in Cold Soils. M.: Izd-vo AN SSSR, 1952, 279. (R)
Darwin C: Cross-fertilization and Self-Fertilization in the Plant World. M.; L.: Selkhozgiz, 1939, 339. (R)
Deeva V.P.: Retardants – Plant Growth Regulators. Minsk: Nayka i tekhnika, 1980, 176. (R)
Dennis D.T., Upper C.D., West C.A.: An enzymic site of inhibition of gibberellin biosynthesis by Amo 1618 and other growth retardants. Plant Physiol., 1965, 40: N 5, 948–952.
De Vries H.: Intracelluläre Pangenesis. Jena, 1889.
De Vries H.: Die Mutationstheorie. Leipzig, 1901, 1903, 1,2.
Devyatnin V.A.: Vitamins. M.: Pishepromizdat, 1948, 227. (R)
Djaparidze L.I.: Leaf respiration in dioecious plants. Soobsh. AN GSSR, 1941, 1: 923–927. (R)
Djaparidze L.I.: New data on the sexually differentiated water content in plants. Soobsh. AN GSSR, 1945a, 6: 37–42. (R)
Djaparidze L.I.: Winter water content in certain woody plants. Soobsh. AN GSSR, 1945b, 6: 205–210. (R)
Djaparidze L.I.: Sexuality in Plants. Tbilisi: Izd-vo AN GSSR, 1963, 307. (R)
Djaparidze L.I.: Sexuality in Plants. Tbilisi: Metsniereba, 1965, 302. (R)
Djaparidze L.I., Kezely T.A.: Differences in the oxidizing properties in the tissues of dioecious plants. Botan. zhurn. SSSR, 1934, 19: 534–538. (R)
Djaparidze L.I., Moniava E.B.: Characteristic features of transpiration in dioecious plants. Soobsh. AN GSSR, 1948, 9: 303–306. (R)
Dobrunov L.G.: Characteristic features of mineral nutrition in hemp. Tr. VNII konopli, 1935a, 8: 85–125. (R)
Dobrunov L.G.: The specific growth and mineral nutrition of male and female hemp plants. Tr. VNII konopli, 1935b, 8: 119–125. (R)

Dressler O.: Zytogenetische Untersuchungen an diploiden und polyploiden Spinat (*Spinacia oleracea*) unter besonderer Berücksichtigung der Geschlechtvererbung àls Grundlage einer Inzuct-Heterosis-Züchtung. Ztschr. Pflanzenzücht., 1958, 40: S. 385–424.

Dressler O.: Erfahrungen bei der Vermethrung und Züchtung monözicher Spinatsorten (*Spinacia oleracea*). Ztschr. Pflanzenzüht., 1973, 70: S. 108–128.

Driga I.E.: Cytological studies of hemp. Tr. VNII konoply, 1934, 5: 65–72. (R)

Dunberg A.: Effect of growth retardants on Norway spruce (*Picea abies*). Physiol. Plant., 1972, 26: N 3, S. 302–305.

Dunberg A.: Gibberellin-like substances from Norway spruce (*Picea abies*). Physiol. Plant., 1973, 28: N2, S. 358–360.

Dunberg O.: Occurrence of gibberellin-like substances in Norway spruce (*Picea abies L. Karst.*) and their possible relation to growth and flowering. Stud. Forest. Succ., 1974, N 111, S. 5–62.

Durand B.: L'expression de sexe chez mercuriales annuelles. Bull. Soc. Franç. Physiol. Végét., 1967, 13: N 2, 195–202.

Durand B., Durand-Rivières R.: Cytokinines et régulation de la synthèse d'une protéine antigénique spécifique de sexe femelle chez une plante dioique, *Mercurialis annua L.* ($2n=16$). C. r. Acad. Sci., 1969, D 269: N 17, 1639–1641.

Durand-Rivières R.: Mise en évidence d'une protéine antigénique spécifique dans les méristèmes et les feuilles femelles de *Mercurialis annua L.* ($2n=16$). C. r. Acad. Sci., 1969, D 268: N 16, 2046–2048.

Dvigubskyi I.A.: Introduction to the Natural History of Plants; Terminology, Best Plant Systems, Anatomy, Physiology, Pathology, and the History of Botany. M., 1823, 12. (R)

Edmond J.B.: Seasonal variation in sex expression of certain cucumber varieties. Proc. Am. soc. Hort. Sci., 1930, 27: 329–331.

Eisuke M., Voshikuzu M., Shigeyuki T.: Studies on the growth behaviour of cucumber in controlled environments. III. Effects of different photoperiodic combinations of various kinds of light on the growth and the sex differentiation of cucumber. J. Jap. Soc. Hort. Sci., 1968, 37: N 4, 328–332.

Eisuke M., Eiji F.: Studies on the photoperiodic sex differentiation in cucumber, *Cucumis sativus L.* IV Effect of light-break. J. Jap. Soc. Hort. Sci., 1970, 39: N 2, 144–148.

Engelbrecht L.: Differences in the development of male and female hemp plants in relation to hormonal regulation. Pr. Inst. sadown. Skiern., 1973, E: N 3, S. 389–397.

Engelsman G.: The influence of light of different spectral regions on the synthesis for photomorphogenesis. IV. Mechanisms of far-red action. Acta bot. Neer., 1968, 17: N 2, blz. 85–89.

Erdelsky K., Herich R.: Prispevok k biochemickej diferenciacii pohlavia konopi (*Cannabis sativa L.*). Biologia, 1956, XI: N 2, 111–115.

Fahn A., Klarman-Kisler N., Ziv D.: The abnormal flower and fruit of May flowering dwarf Cavendish bananas. Bot. Gaz., 1961, 123: 116–125.

Fedin M.A., Gyska M.N.: The gametoside-like effect of Ethrel on aestival wheat. Khimia v sel. khoz-ve, 1975, 13: 41–42. (R)

Fisch C.: Über Zahlenverhältnisse der Geschlechter beim Hanf. Ber. Dtsch. Bot. Ges., 1920, 5: S. 136–142.

Frankel R., Galun E.: Pollination mechanisms. Reproduction and plant breeding. B. New York, Springer-Verlag, 1977, S. 135.

Freytag A.H., Lira E.P., Isleib D.R.: Cucumber sex expression modified by growth regulators. Hort. Sci., 1970, 5: N 6, 509–511.

Friedlander M., Atsmon D., Galun E.: Sexual differentiation in cucumber: the effects of abscisic acid and other growth regulators on various sex genotypes. Plant Cell Physiol., 1977, 18: N 1, 261–269.

Fuchs Y., Lieberman M.: Effects of kinetin, IAA and gibberellin acid on ethylene production and their interactions in growth of seedlings. Plant. Physiol., 1968, 43: N 12, 2029–2036.

Fuchs E., Atsmon D., Halevy A.H.: Adventitious staminate flower formation in gibberellin treated gynoecious cucumber plants. Plant Cell Physiol., 1977, 18: N 6, 1193–1201.

Fujii T.: Radiation effect upon sex expression in cucumber. Annu. Rep. Nat. Inst. Genet. Jap., 1972 (1973), 23: 76–77.

Fursa T.: Pollination in watermelon. Kartofel i ovoshi, 1969, 8: 35–36. (R)

Galialo M.A., Vladimirova G.E., Vinogradov A.P., et al.: On the specificity of the chemical reaction of Dr. Manoylov. Vracheb. gaz., 1926, 13: 645–650. (R)

Galinat W.C., Naylor A.W.: Relation of photoperiod to inflorescence proliferation in *Zea mays L.* Am. J. Bot., 1951, 38: 38–42.

Galun E.: Effect of gibberellic acid and naphthalene acetic acid on sex expression and some morphological characters in the cucumber plants. Phyton., 1959a, 13: N 1, S. 1–8.

Galun E.: The role of auxin in the expression of the cucumber. Physiol. Plant., 1959b, 12: N 1, S. 48–61.

Galun E.: Study of the inheritance of sex expression in the cucumber. The interaction of major genes with modifying genetic and non-genetic factors. Genetica, 1961, 32: N 1/2, blz. 134–163.

Galun E., Jung Y., Lang A.: Culture and modification of male cucumber buds in vitro. Nature (London), 1962, 194: N 4828, 596–598.

Galun E., Izbar S., Atsmon D.: Determination of relative auxin content in hermaphroditic and andromonecious *Cucumis sativus L.* Plant Physiol., 1965, 40: 321–326.

Gamburg K.Z.: Biochemistry of Auxin and its Effect on Plant Cells. Novosibirsk: Nauka, 1976, 272. (R)

Gardner V.R.: Studies in the nutrition of the strawberry. Res. Bull. Agr. Exp. Stat. Univ. Mo., 1922, 57: 1–32. (R)

Gardner V.R., Bradford F.Ch., Hooker G.D.: Basic Plant Breeding. M.; L.: Selkhozgiz, 1934, 404. (R)

Garner W.W., Allard H.A.: Effect of the relative length of day and night and other factors of the environment on growth and reproduction in plants. J. Agr. Res., 1920, 18: N 11, 553–606.

Gaspar T., Smith D., Thorpe T.: Argument supplémentaire en Faveur d'une variation inverse du niveau auxinique endogène au cours des deux premières phases de la rhizogenèse. C. r. Acad. Sci., 1977, D 235: N 4, 327–330.

Geddes P., Thompson J.A.: The Evolution of Sex. London, 1889, p. 97.

Ghosh M.S., Bose T.K.: Sex modification in cucurbitaceous plants by using CCC. Phyton, 1970, 27: N 2, S. 131–135.

Giard A.: Les variations de la sexualité chez les végétaux. C. r. Soc. Biol. (Paris), 1898, 10: 730–732.

Gogina E.E.: On the difference in seed production between bisexual and female Thymus plants. Bull. Gl. botan. sada AN SSSR, 1971, vyp.82, 72–76. (R)

Goldschmidt R.: Vorläufige Mitteilung über weitere Versuche zur Vererbung und Bestimmung des Geschlechts. Biol. Zentr.-Bl., 1915, 35: S. 565.

Goldschmidt R.: Untersuchungen über Intersexualität. Ztschr. indukt. Abstammungs- und Vererbungslehre, 1920, 23: S. 1.

Goldschmidt R.: Physiologische Theorie der Vererbung. B., 1927, S. 48.

Gorshkov P.A., Sazhko M.M.: The influence of nitrogen nutrition on the expression of monoecism in hemp and on the formation of different sex types in monoecious hemp. Agrokhimia, 1964, 7: 23–24. (R)

Gregory F.G., Purvis O.N.: Abnormal flower development in barley involving sex reversal. Nature (London), 1947, 160: 221–222.

Grishko N.N.: The problem of sex in hemp. Tr. VNII konopli. Kiev. 1935, 8: 197–241. (R)

Grossgeim A.A.: On the graphic representation of the systems in flowering plants. Sov. botanika 1945, 13: 3–27. (R)

Grunberg O.I.: The determination of sex in dioecious plants using E.O. Manoylov's reaction. Vracheb. gaz., 1924a, 5: 108. (R)

Grunberg O.I.: Supplement to Dr. Manoylov's work: "The determination of sex in dioecious plants using a chemical reaction." Tr. po prikl. botanike, genetike i selekcii, 1924b, 13: 506. (R)

Hall W.C.: Effects of photoperiod and nitrogen supply on growth and reproduction in the gherkin. Plant Physiol., 1949, 24: 753–755.

Halsted B.D.: Experiments with hemp. Rep. N. J. Agr. Res. Stat., 1901, 21: 455–457.

Hartmann C., Durand B.: Sur les activités respiratoires et photosynthétiques de *Mercurialis annua L.* (2n=16), espèce dioique. C. r. Acad. Sci., 1969, D 269: N 18, 1762–1763.

Hashizume H.: The effect of gibberellin upon flower formation in *Cryptomeria japonica*. J. Jap. Forest. Soc., 1959a, 41: N 10, 375–381.

Hashizume H.: The effect of gibberellin upon flower formation and sex transition to female in *Chamaecyparis obtusa* and *Ch. Lowsoniana*. J. Jap. Forest. Soc., 1959b, 41: N 11, 458–463.

Hashizume H.: The germination of *Chamaecyparis obtusa* seeds collected from cones borne by spraying with gibberellin. J. Jap. Forest. Soc., 1960, 42: N 5, 176–180.

Hashizume H.: The effect of gibberellin in rellin upon sex differentiation in *Cryptomeria japonica* strobiles. III. Changes of endogenous growth substances, carbohydrates and nitrogen in new shoots in relation to flower induction by gibberellin. J. Jap. Forest. Soc., 1961, 43: N 3, 47–50.

Hashizume I., Iizuka M.: Induction of female organs in male flowers of Vitis species by zeatin and dihydrozeatin. Phytochemistry, 1971, 10: N 12, 2653–2655.

Hayaschi F., Boerner D.R., Peterson C.E., Sell M.H.: The relative content of gibberellin in seedlings of gynoecious and monoecious cucumber (*Cucumis sativus*). Phytochemistry, 1971, 10: N 1, 57–62.

Heftmann E.: Functions of sterols in plants. Lipids, 1971, 6: N 2, 128–133.

Heftmann E.: Biochemistry of Steroids. M.: Mir, 1972, 175. (R)

Heide O.M.: Environmental control of sex expression in Begonia. Plant. Physiol., 1969, 61: 279–285.

Heide O.M.: Circadian rhythmicity in photoperiodic regulation of regeneration in Begonia leaves. Physiol. Plant., 1978, 43: S. 266–270.

Heinz A.: Cormofyternas Fylogenis. Lund, 1927, S. 105.

Hemberg T., Westlin P.E.: The quantitative yield in purification of cytokinins; model experiments with kinetin, 6-furfurylaminopurine. Physiol. Plant., 1973, 28: S. 228–230.

Hemphill D.D. jun., Baler L.R., Sell H.M.: Different sex phenotypes of *Cucumis sativus*

L. and *C. melo L.* and their endogenous gibberellin activity. Euphytica, 1972, 21: blz. 285–291.

Henson I.E., Wareing P.F.: Cytokinins in *Xanthium strumarium L.*: Distribution in the plant and production in the root system. J. Exp. Bot., 1976, 27: N 101, 1268–1278.

Herich R.: Gibberellin and sex differentiation of flowering plants. Nature (England), 1960, 188: N 4750, 599–600.

Herich R.: Gibberellinsäure und Geschlechtsdifferentiation der Pflanzen. Acta Fac. rerum natur. Univ. comen. Bot., 1961, 5: N 11/12, S. 627–633.

Herich R., Priehradny S.: Prispevok k biochemickej diferenciacii pohlovia konopi (*Cannabis sativa L.*). Biologia, 1955, 10: N 3, S. 346–350.

Herskowitz I.: Genetics. Boston, Toronto: Little Brown and Co., 1965, p. 671.

Herskowitz I.: Genetics. M.: Nauka, 1968, 701. (R)

Hertel R., Thompson K.S., Russo V.E.A.: In vitro auxin binding to particulate cell fractions from corn coleoptiles. Planta, 1972, 107: N 4, 325–340.

Heslop-Harrison J.: Auxin and sexuality in *Cannabis sativa*. Physiol. Plant., 1956, 9: N 4, S. 588–597.

Heslop-Harrison J.: The experimental modification of sex expression of flowering plants. Biol. Rev., 1957, 32: 38–90.

Heslop-Harrison J.: Growth substances and flower morphogenesis. J. Linnean Soc. London, 1959, 56: 269–281.

Heslop-Harrison J.: The experimental control of sexuality and inflorescence structure in *Zea mays L.* Proc. Linnean Soc. London, 1961, 172: 108–123.

Heslop-Harrison J.: Sex expression in flowering plants. Meristem and differentiation. Brookhaven Symp. Biol., 1963, 16: 109–125.

Heslop-Harrison J.: The control of flower differentiation and sex expression. In: Régulateurs Naturels de la Croissance Végétale: (Fifth Intern. Conf. on Plant Growth Substances). Paris: C. N. R. S., 1964, 597–609.

Heslop-Harrison J.: Sexuality in angiosperms. In: Plant Physiology—A treatise/Ed. Steward F. C. N. Y.: New York, Academic Press, 1972, 6 (C): 133–289.

Heslop-Harrison J., Heslop-Harrison Y.: Studies of flowering plant growth and organogenesis. I. Morphogenetic effects of 2,3,5 triiodobenzoic acid on *Cannabis sativa*. Proc. Roy. Soc. Edinburgh B. (Biology), 1956/1957, 66: N 4, 409–423.

Heslop-Harrison J., Heslop-Harrison I.: Studies on flowering plant growth and organogenesis. II. The modification of sex expression in *Cannabis sativa* by carbon monoxide. Proc. Roy. Soc. Edinburgh B (Biology), 1956/1957, 66: N 4, 424–434.

Heslop-Harrison J., Heslop-Harrison Y.: Photoperiod, auxin and balance in a long-day plant. Nature, 1958a, 181: 100–102.

Heslop-Harrison J., Heslop-Harrison Y.: Long-day and auxin induced male sterility in *Silene pendula L.* Port. Acta Biol., 1958b, A5: N 2, 79–94.

Hewitt E.W., Wareing P.F.: Cytokinins in *Populus robusta*: changes during chilling and bud burst. Physiol. Plant., 1973, 28: N 3, S. 393–399.

Heyer F.: Untersuchungen über das Verhältnis des Geschlechts bei einhäusigen und zweihäusigen Pflanzen. Ber. Landw. Inst. Halle, 1884, S. 5–81.

Higgins T.Y.V., Zwar J.A., Jacobsen J.V.: Gibberellic acid enhances the level of translatable mRNA for α-amylase in barley aleurone layers. Nature, 1976, 260: N 55–57, 166–169.

Hirata K.: Sex determination in hemp. Jap. J. Genetics, 1927, 19: N 1, 82–85.

Hoffmann H.: Über Sexualität. Bot. J., 1885, 43: N 145, 161–165.

Hoffmann W.: Gleichzeitig reifender. Züchter, 1941, 13: S. 277–283.

Hoffmann W.: Die Vererbung der Geschlechtsformen des Hanfes (*Cannabis sativa*). II Züchter, 1952, 22: S. 147–158.
Howlett F.S.: The effect of carbohydrate and of nitrogen deficiency upon microsporogenesis and the development of the male gametophyte in the tomato, *Lycopersicum esculentum Mill.* Ann. Bot., 1936, 50: 767–769.
Howlett F.S.: The modification of flower structure by environment in varieties of *Lycopersicum esculentum*. J. Agr. Res., 1939, 58: 79–117.
Hughes W.G., Bennet M.D., Bodden J.J., Galunopoulon S.: Effects of time of application of ethrel on male sterility and ear emergence in wheat *Triticum aestivum*. Ann. Appl. Biol., 1974, 76: 243–252.
Hutchinson J.: The Families of Flowering Plants. L., 1926–1934, 1–2: 2nd ed., 1959.
Hylmö B.: Einwirkung durch Hormonbehandlung auf das Geschlecht der Spinacia. Bot. Notiser, 1941, S. 389–391.
Ivanov N.N.: Specific Differences between Proteinaceous Substances. M.; L.: Snabkoopgiz, 1931, 72. (R)
Ivanov N.N.: Biochemical differences between the sexes. Uspekhy biol khimii, 1935, 11: 164–202. (R)
Ivanova I.: The interactions between gibberellin, auxin, and natural inhibitors in regulating plant elongation. Fiziol. na rasteniata, 1974, 2: 257–268.
Ivanova Y.A.: Influence of gibberellic acid and of the CCC retardant on the content of auxins and inhibitors in hemp (*Cannabis sativa L.*). C. r. Acad. Bulg. Sci., 1971, 24: N 10, 1407–1410.
Ivonis I.U., Kyalina L.V., Khokhlina E.V.: The gibberellin-like substances in the needle of Norway spruce cones that bear a high number of microstrobili. Fiziol. rastenyi, 1974, 2: 257–268. (R)
Iwahori S., Lyons J.M., Sims W.L.: Induced femaleness in cucumber by 2-chloroethanephosphonic acid. Nature, 1969, 222: N 5190, 271–272.
Iwahori S., Lyons J.M., Smith O.E.: Sex expression in cucumber plants as affected by 2-chloroethylphosphonic acid, ethylene and growth regulators. Plant Physiol., 1970, 46: 412–415.
Jaiswal V.S., Mohan Ram H.Y.: Inhibition of GA_3 induced extension growth and male flower formation in female plants of *Cannabis sativa* by cycloheximide. Curr. Sci. (India), 1974, 43: N 24, 800–801.
Janick J., Iizuka M.: Sex determination in spinach. In: 19th Intern., Hort. Congr. Brussels, 1962, 82–88.
Jones D.F.: Sex intergrades in dioecious maize. Am. J. Bot., 1939, 26: N 8, 412–415.
Jones K.L.: Studies in Ambrosia. IV Effects of short photoperiod and temperature on sex expression. Am. J. Bot., 1947, 34: N 7, 317–377.
Jong A.W. de, Smit A.L., Bruinsma J.: Pistil development in Cleome flowers. II. Effects of nutrients on flower buds of *Cleome iberidella* grown in vitro. Ztschr. Pflanzenphysiol., 1974, 72: N 3, S. 227–236.
Jong A.W. de, Bruinsma J.: Pistil development in Cleome flowers. III. Effects of growth regulating substances on flower buds of *Cleome iberidella* grown in vitro. Ztschr. Pflanzenphysiol., 1974a, 73: N 2, S. 142–151.
Jong A.W. de, Bruinsma J.: Pistil development in Cleome flowers. IV. Effects of growth-regulating substances on female abortion in *Cleome spinosa Jasq.* Ztschr. Pflanzenphysiol., 1974b, 73: N 2, S. 152–159.
Joung J.J., Knights B.A., Hillman J.K.: The metabolism of estrogens in vivo and in vitro by *Phaseolus vulgaris*. Ztschr Pflanzenphysiol., 1979, 94: N 4, S. 307–316.
Joyet-Lavergne Ph.: La physico-chimie de la sexualité. Protoplasma. P., 1931, p. 5.

Kalantyr M.S.: The regeneration of seed fertility in Eucommia. Agrobiologia, 1947, 2: 87–97. (R)

Kalyagin V.N.: The effect of gibberellin on sex expression in pumpkin. Bul. VIR, 1973, 29: 105–113. (R)

Kandina G.V.: The effects of treating sprouting cucumber seed with cold temperatures. Izv. Mold. fil. AN SSSR, 1958, 5: 85–94. (R)

Karchi A.Z.: Effects of 2-chloroethanephosphonic acid on flower types and flowering sequences in muskmelon. J. Am. Soc. Hort. Sci., 1970, 95: N 5, 515–518.

Kardo-Sysoeva E.O.: Gynandromorphism in *Salix cineraea L.* Tr. Leningr. o-va estestvoispytateley, 1924, 54: 41–44. (R)

Kaushik M.P., Bisaria A.K.: Combined effect of some growth regulators and day length on sex expression in muskmelon. Ind. J. Exp. Biol., 1974, 12: N 1, 111–112.

Kefeli V.I.: Growth of Plants. M.: Kolos, 1973, 120. (R)

Kefeli V.I.: Natural Growth Inhibitors and Phytohormones. M.: Nauka, 1974, 253.

Kefeli V.I., Turetskaya R.Kh.: Determination of free auxins and inhibitors in plant tissues. In: Methods of Identifying Growth Regulators and Herbicides. M.: Nauka, 1966, 20–44. (R)

Kefeli V.I., Turetskaya R.Kh., Kof E.M., et al.: Determination of the biological activities of free auxins and growth inhibitors in plant material. In: Methods of Identifying Phytohormones, Growth Inhibitors, Defoliants, and Herbicides. M.: Nauka, 1973, 7–21. (R)

Kefeli V.I., Yanina L.I., Zhloba N.M.: The use of *Amaranthus caudatus L.* cotyledons in determining cytokinin activity. Ukr. botan. zhurn., 1977, 34: 643. (R)

Kende H., Gardner G.: Hormone binding in plants. Ann. Rev. Plant Physiol., 1976, 27: 267–290.

Kender W.J., Ramaily G.: Regulation of sex expression and seed development in grapes with 2-chloroethylphosphonic acid. Hort. Science, 1970, 5: 491–492.

Kezeli T.A.: Changes in catalase and peroxidase activities in willow related to age and root formation. Soobsh. AN GSSR, 1942, 3: 353–357. (R)

Kezeli T.A.: Changes in peroxidase activity in certain dioecious plants. Soobsh. AN GSSR, 1944, 5: 279–284. (R)

Kezeli T.A., Djaparidze L.I., Tarasashvili K.M.: The dynamics of vitamin C in persimmon (*Diospyros lotus L.*). Soobsh. AN GSSR, 1945, 6: 281–285. (R)

Kezeli T.A., Tarasashvili K.M.: Changes in ascorbic acid content in certain dioecious plants. Soobsh. AN GSSR, 1946, 7: 57–60. (R)

Kholodnyi N.G.: New evidence supporting the hormonal theory of tropisms. Zhurn. Rus. botan. o-va, 1928, 13: 191. (R)

Kholodnyi N.G., Fitogormony: On the physiology of hormonal changes in plants. Kiev: Izd-vo AN USSR, 1939, 265. (R)

Khokhlov S.S.: Plants Bearing Parthenocarpic Seeds. Uchen. zap. Sarat. gos. un-ta, vyp. 1 (Biologia), 1946, 16: 3–74. (R)

Khokhlov S.S.: On the Evolution of Higher Plants. Saratov: Sarat. gos. ped. in-t, 1949, 196.

Khrianin V.N.: The effects of extra-radical introduction of nutrients and growth regulators on the physiological processes, plant productivity, and overall quality of hemp (*Cannabis sativa*). Uchen. zap. MOPI im. N.K. Krupskoy, 1964, 153: 323–345.

Khrianin V.N.: The effects of growth regulators on productivity and quality of hemp. Lion i konoplya, 1965, 6: 33–34. (R)

Khrianin V.N.: The effect of gibberellin on the anatomy of the stem in hemp. Bul. Gl. botan. sada, 1966a, 62: 103–106. (R)

Khrianin V.N.: The effect of gibberellin on the productivity and quality of fiber in hemp. Khimia v sel. Khoz-ve, 1966b, 6: 33–34. (R)

Khrianin V.N.: Effects of Gibberellin on Physiological Processes and Crop Yields in Hemp (*Cannabis sativa*). Avtoref. dis. kand. biol. nauk. 14: MOPI im. N.K. Krupskoy, 1967, 16. (R)

Khrianin V.N.: Sex transformation in hemp as a result of gibberellin treatment. S.-kh. biologia, 1969, 4: 753–758. (R)

Khrianin V.N.: The possible uses of gibberellin and gibberellic acid in hemp production. Fiziol. rastenyi, 1971, 18: 638–641. (R)

Khrianin V.N.: The content of gibberellins, auxins, and inhibitors in the seeds and seedlings in hemp. Biol. nauki, 1974, 7: 89–92. (R)

Khrianin V.N.: The content of endogenous regulators in the leaves of hemp. Biol. nauki, 1975, 5: 76–79. (R)

Khrianin V.N.: The changes of natural regulators in the process of hemp ontogenesis. In: Plant Growth Regulators: Proc. II. Intern. Symp. on Plant Growth regulators. Sofia, Bulgaria, October 1975/Ed. T. Kudrev, I. Ivanova, E. Karanov. Sofia: Publ. House of the Bulg. Acad. Sci., 1977, pp. 270–285. (R)

Khrianin V.N.: Sex transformation in plants as a result of gibberellin and chlorocholine-chloride treatments. Dokl. VASKhNIL, 1978, 3: 81–21. (R)

Khrianin V.N., Chailakhyan M.Kh.: The influence of daylength of sex expression in hemp and spinach. Dokl. AN ArmSSR, 1977, 55: 186–191. (R)

Khrianin V.N., Milyaeva E.L.: The influence of gibberellin on the differentiation of stem apices in hemp. Dokl. AN SSSR, 1977, 234: 982–984. (R)

Khrianin V.N., Chailakhyan M.Kh.: The biological activities of cytokinins and gibberellins present in the leaves during sex expression in dioecious plants. Fiziol. rastenyi, 1979, 26: 1008–1015. (R)

Kizel A.R.: Chemistry of Protoplasm. M.; L.: Izd-vo AN SSSR, 1940, 624. (R)

Kizel A., Pashkevich O.: Differences in the amino-acid composition of proteins in the leaves of male and female hemp plants. Biokhimia, 1937, 11: 666–673. (R)

Knott J.E.: Rapidity of response of spinach to change in photoperiod. Plant Physiol., 1932, 7: 127–129.

Knott J.E.: Effect of localized photoperiod on spinach. Proc. Am. Soc. Hort. Sci., 1934, 31: 152–155.

Köhler D.: Geschlechtsbestimmung bei Blütenpflanzen. Ergeb. Biol., 1964a, 27: S. 98–115.

Köhler D.: Veränderung des Geschlechts von Cannabis sativa durch Gibberellinsäure. Ber. Dtsch. bot. Ges., 1964b, 78: N 7, S. 275–281.

Köhler F.: Beitrag zur Kenntnis der Sexualreaktionen von *Mucor mucedo (Bref)*. Planta, 1934, 23: N 3, S. 358–377.

Kölreuter J.: Vorläufige Nachricht von einigen das Geschlecht der Pflanzen betreffenden Versuchen und Beobachtungen. Leipzig, 1961, S. 65.

Kölreuter J.: Treatise on Sexuality and Hybridization of Plants. M.; L.: Ogiz-Selkhozgiz, 1940, 212.

Komarov V.L.: Treatise on Plant Species. M.; L.: Izd-vo AN SSSR, 1940, 212. (R)

Konopskaya L.N.: Cytokinin content in sprouting seeds of pea. Dokl. AN SSSR, 1977, 236: 1270–1272. (R)

Konstantinova T.N., Bavrina T.V., Aksenova N.P.: Characteristic features of the photoregulation of generative morphogenesis *in vivo* and *in vitro*. In: Photoregulation of Generative Metabolism and Plant Morphogenesis. M.: Nauka, 1975, 186–198. (R)

Kopcewicz J.: Estrogens in developing bean (*Phaseolus vulgarus*) plants. Phytochemistry, 1971, 10: N 7, 1423–1427.

Kopcewicz J.: Estrogens in the short-day plants *Perilla ocimoides* and *Chenopodium rubrum* grown under inductive and non-inductive light conditions. Ztschr. Pflanzenphysiol., 1972a, 67: N 4, S. 373–376.

Kopcewicz J.: Influence of indole-3-acetic acid, kinetin and abscisic acid on the estrogen content of beans. Biol. Plant. Acad. Sci. Bohemosl., 1972b, 14: N 3, S. 223–226.

Kopcewicz J., Chrominski A.: Estrogens in 2-chloroethylphosphonic acid induced femaleness of *Cucurbita pepo L.* Experientia, 1972, 28: N 5, 603–604.

Kopcewicz J., Rogozinska J.H.: Effect of estrogens and gibberellic acid on cytokinin and abscisic acid-like compound contents in pea. Experientia, 1972, 28: 1516–1517.

Kotaeva D.V.: Sex-related differences in nucleic acid content of leaves in certain dioecious deciduous plants. Soobsh. AN GSSR, 1973, 70: 693–696. (R)

Kotaeva D.V.: Seasonal changes in nucleic acids in dioecious plants: Avtoref. dis. ... kand. biol. nauk. Tbilisi: Tbil. gos. un-t, 1975, 35. (R)

Kovalchuk U.G.: Biogenous ethylene and the chemical interactions in plants. In: Interactions of Plants and Microorganisms in Plant Communities. Kiev: Nauk. dumka, 1977, 12–20. (R)

Kolaleva L.V., Khrianin V.N., Chailakhyan M.Kh.: Specificity of proteins in the process of sex expression in hemp. Dokl. AN SSSR, 1980, 252: 1275–1276. (R)

Kozo-Polyanskyi B.M.: Teratology of the flower: new theoretical questions. Sov. botanika, 1937, 6: 56–70. (R)

Kozo-Polyanskyi B.M.: The importance of different methods in plant systematics. Probl. botaniki, 1950, 1: 28–69. (R)

Kravyazh K., Karavayko N.N., Kof E.M., et al.: Interaction between abscisic acid and cytokinin in the regulation of the growth and viridescence of pumpkin cotyledons. Fiziol. rastenyi, 1977, 24: 160–167. (R)

Krechetovich L.M.: On the Evolution of the Plant World. M.: MOIP, 1952, p. 1351. (R)

Krekule J., Seidlova F.: Effects of exogenous cytokinins on flowering of the shortday plant *Chenopodium rubrum L.* Biol. Plant., 1977, 19: 142–149.

Krishnamoorthy H.N.: Effect of GA_3, GA_{4+7} and GA_5 on the sex expression of *Luffa acutangula* var. H-2. Plant. Cell Physiol., 1972, 1: 381–383.

Krishnamoorthy H.N., Talukdar A.R.: Chemical control of sex expression in *Zea mays L.* Ztschr. Pflanzen. Physiol., 1976, 79: S. 91–94.

Kroker V.: Growth of Plants. M.: Izd-vo inostr. lit., 1950, 149–155. (R)

Kubarev P.I.: Difference in nucleic acid content between male and female corn flower clusters. Fiziol. rastenyi, 1965, 12: 968–970. (R)

Kubarev P.I.: Sexual types in plant ontogenesis: Avtoref. dis. ... kand. biol. nauk. L.: VIR, 1966, p. 20. (R)

Kulaeva O.N.: The influence of roots on leaf metabolism in relation to the effect of kinetin on the leaves. Fiziol. rastenyi, 1962, 9: 229–239. (R)

Kulaeva O.N.: Cytokinins, Structure and Function. M.: Nauka, 1973, p. 264. (R)

Kulaeva O.N.: The role of phytohormones in regulating protein synthesis in plants. In: Plant Proteins and their Biosynthesis. M.: Nauka, 1975, pp. 220–234. (R)

Kulaeva O.N.: The mode of action of cytokinins. In: Growth of Plants and Natural Regulators. M.: Nauka, 1977, pp. 216–234. (R)

Kumarasamy S.: Ethrel—the versatile plant hormone. Pesticides, 1972, 6: N 10, 25–27.

Kun R.: Substances that stimulate fertilization and determine the sex in plants and animals. Uspekhi sovrem. biologii, 1941, 14: 112–120. (R)

Kuperman F.M., Rzhanova E.I., Kapitonova T.A.: The main stages in the formation of fruit bearing organs in corn. In: Stages in the Formation of Generative Organs in Grasses. M.: Izd-vo MGU, 1955: 278–316. (R)

Kuptsov A.I.: A unisexual female sunflower. Sots. rastenievodstvo, 1935, 14: 149–150. (R)

Kurilova A.V.: The transpirational coefficient and the influence of soil humidity on the development and productivity of hemp. Botan. zhurn., 1935, 20: 46–51. (R)

Kursanov A.L.: Interactions between plant physiological processes. XX Timiryazevskie chtenia. M.: Izd-vo AN SSSR, 1960, p. 44. (R)

Kursanov A.L.: Transport of Assimilates in Plants. M.: Nauka, 1976, p. 646. (R)

Kutuzova S.N.: Influence of gibberellin on the growth, development, and sexuality of hemp. Avtoref. dis. . . . kand. biol. nauk. L.: LGU, 1969, 24. (R)

Kuznetsov A.I., Andreev V.S., Belyaeva R.G., et al.: Genetic control of hormonal regulation of sex formation in *Papaver somniferum*. In: Proceedings of the 1st All-Union Conference on the Genetics of Plant Development. Tez. dokl. 1-vo Vsesoyuz. sovesh. "Genetika razvitia rastenyi." Tashkent: 1980, pp. 104–105. (R)

Lacombe J.-P.: Discrimination des sexes on fonction de caractères végétatifs précoces chez le Chanvre dioique (*Cannabis sativa L.*). Physiol. végét., 1980, 18: N 3, 419–430.

Laibach F., Kribben F.J.: Der Einfluss von Wuchsstoff auf die Bildung männlicher und weiblicher Blüten bei einer monözischen Pflanze (*Cucumis sativa L.*). Ber. Dtsch. Bot. Ges., 1950a, 62: S. 53–55.

Laibach F., Kribben F.J.: Der Einfluss von Wuchsstoff auf das Geschlecht der Blüten bei einer monözischen Pflanzen. Beitr. Biol. Pflanzen, 1950b, 28: S. 64–67.

Laibach F., Kribben F.J.: Die Bedeutung des Wuchsstoffs für die Bildung und Geschlechtsbestimmung der Blüten. Beitr. Biol. Pflanzen, 1951, 28: S. 131–144.

Lang A.: The effect of gibberellin upon flower formation. Proc. Nat. Acad. Sci. USA, 1957, 43: N 8, 709–717.

Lang A.: Factors affecting sex changes in the flowers of *Carica papaya*. Proc. Am. Soc. Hort. Sci., 1961, 77: 252–264.

Lang A.: Stem elongation in a rosette plant induced by gibberellic acid. Naturwissenschaften, 1965a, 43: N 11, S. 257–258.

Lang A.: Induction of flower formation on Biennial Hyoscyamus by treatment with gibberellin. Naturwissenschaften, 1965b, 43: N 12, 284–285.

Law J., Stoskopf N.C.: Further observations on ethephon (Ethrel) as a tool for developing hybrid cereals. Can. J. Plant. Sci., 1973, 53: 765–766.

Lebedev S.I.: The content of carotene in pollen and its effect on the growth of pollen tubes. Dokl. AN SSSR, 1948, 59, 5: 987–990. (R)

Lebedeva A.T., Yurina O.V.: The effect of greenhouse selection for strong root system in cucumber plants on the production and weight of seed. Tr. molodykh uchionykh i aspirantov po selektsii i semenovodstvu ovoshnykh kultur. L.: VIR, 1970, 3: 50–54. (R)

Leisle F.F., Makarova N.A.: The effect of soil humidity on the growth, development and other processes in the dioecious plant *Melandrium album Garcke*. Tr. Botan. in-ta AN SSSR, 1950, 7: 221–237. (R)

Letham D.S.: Chemistry and physiology of kinetin-like compounds. Ann. Rev. Plant Physiol., 1967, 18: 349–351.

Letham D.S.: Regulators of cell division in plant tissues. XXI. Distribution coefficients for cytokinins. Planta, 1974, 118: N 5, S. 361–363.

Letham D.S., Williams M.W.: Regulators of cell division in plant tissues. VIII. The cytokinins of the apple fruit. Physiol. Plant., 1965, 22: 925–928.

Levchenko V.I.: Changes in the morphology of hemp flowers resulting from short day and traumatical injuries. Tr. VNII konopli, 1937, 5: 109–124. (R)

Levitskyi G.A.: Natural and induced changes in the flower structure in *Veratrum nigrum* L. Tr. po prikl. botanike, genetike i selektsii, 1925, 14: 97–112. (R)

Lieberman M.: Biosynthesis and regulatory control of ethylene in fruit ripening. Rev. Physiol. Veg., 1975, 13: 489–499.

Lieberman M., Kunishi A.T.: Thoughts on the role of ethylene in plant growth and development. In: Plant Growth Substances/Ed. D.J. Carr. New York, Springer-Verlag, 1972, pp. 549–560.

Lilienfeld H.: Die Resultäte einiger Bestäubungen mit verschiedenaltrigen Pollen bei *Cannabis sativa*. Biol. Zentr.-Bl., 1921, 41: S. 296–303.

Linnaei C.: Praeludio Sponsaliorum Plantarum. Upsal, 1729.

Linnaei C.: Species Plantarum, exhibens Plantas rite cognitas, ad genera relatas, cum differentiis specificis, nominibus trivialibus, synonymis, selectis, locis natalibus, secundum Systema sexuale digestas. Holmiae, 1753.

Linnaei C.: Systema Naturae, sive Regna tria Naturae systematice proposita per classes, ordines, genera et species. Lugduni Batavorum, 1767, 2: S. 22–25.

Loewe S., Spohr E.: Nachweis und Gehaltsbestimmung des weiblichen Brunsthormons in weiblichen Organen des Pflanzenreiches. S. B. Akad. Wiss., Wien, 1926, 63: S. 167–169.

Löve A., Löve D.: Experimental sex reversal in plants. Svensk bot. tidskr., 1940, 34: 248–254.

Löve A., Löve D.: Experiments on the effects of animal sex hormones on dioecious plants. Ark Bot., 1945, 32a: 1–32.

Loy J.B.: Effects of (2-chloroethyl) phosphonic acid and succinic acid–2,2-dimethylhydrazide on sex expression in muskmelon. J. Am. Soc. Hort. Sci., 1971, 96: 641–642.

Lozhnikova V.N.: Natural gibberellins and their significance in photoperiodism and vernalization. Avtoref. dis. Kand. Biol. Nauk. M.: IFR, 1965: 21. (R)

Lozhnikova V.N., Khlopenkova L.P., Chailakhyan M.Kh.: A method to identify natural gibberellins in plant tissues. Agrokhimia, 1967, 10: 132–135. (R)

Lozhnikova V.N., Khlopenkova L.P., Chailakhyan M.Kh.: Identification of natural gibberellins in plant tissues. In: Methods of identifying phytohormones, growth inhibitors, defoliants, and herbicides. M.: Nauka, 1973, pp. 50–58. (R)

Lozhnikova V.N., Chailakhyan M.Kh.: Reaction to the interruption of darkness by light and growth inhibitors. Dokl. AN SSSR, 1975, 222: 1242–1245. (R)

Lubich F.P.: Sex expression in hemp as a function of the location of the fruit in the flower cluster. Agrobiologia, 1950, 1: 153–155. (R)

Lubochinsky B., Zalta J.P.: Microdozage colorimétrique de l'azote ammoniaca. Bull. Soc. Chim. Biol., 1954, 36: 1363–1365.

Lushinskyi V.V.: Comparative physiological survey of different ecological types of hemp. Tr. po prikl. botanike, genetike i selektsii, 1963, 35: 204–210. (R)

Lvova I.N.: Characteristics of flower formation in the *Cucurbitaceae*; an evolutionary perspective. In: Proceedings of the General Biology Conference, Dedicated to the 110th Anniversary of Darwinism. Tomsk, 1959, pp. 128–130. (R)

Lvova I.N.: Sexuality in Plants. M.: Izd-vo MGU, 1963a, p. 56. (R)

Lvova I.N.: The formation of generative organs in *Cucurbitaceae* in different spectral compositions of light. In: Experimental Morphogenesis. M.: Izd-vo MGU, 1963b, pp. 118–122. (R)

Lvova I.N.: Biological control by treatment of cucumber with physiologically active substances. M.: Izd-vo MGU, 1971, p. 6. (R)

Lvova I.N.: Effects of physiologically active substances on the sex of cucumber flowers: sexual processes and embryogenesis in plants. M.: Izd-vo MGU, 1973, pp. 146–147. (R)

Lvova I.N., Bakhanova S.G.: The effects of quality of light on growth and development of cucumber. In: Plant Morphogenesis. M.: Izd-vo MGU, 1961, 2: 79–83. (R)

Lvova I.N., Bakhanova S.G.: Biological control of growth and development of melon: biological control in agriculture. M.: Izd-vo MGU, 1962, p. 38. (R)

Maekawa T.: On the phenomena of sex transition in *Arisaema japonica* Bl. J. Coll. Agr. Hokkaido Imp. Univ., 1924, 13: 217–305.

Maekawa T.: On intersexualism in *Arisaema japonica*. Jap. J. Bot., 1927, 3: 205–207.

Maekawa T.: Widerstands und Selbstregulierungsvermögen gegen Geschlechtsänderung bei Hanefplanzen und seine Bezichung zur Theorie der Geschlechtsbestimmung. Jb. Wiss. Bot., 1929, 70: N 4, S. 512–564.

Magruder R., Allard H.A.: The effect of controlled photoperiod on the production of seed stalks in eight varieties of spinach. Proc. Am. Soc. Hort. Sci., 1937, 34: 502–505.

Makarevich V.A.: The morphological structure of hemp. Tr. VNII konopli, 1935, 8: 23–44. (R)

Makarevich V.A.: The anatomical structure of the stem in hemp. Tr. VNII konopli, 1935a, 8: 45–56. (R)

Makarevich V.A.: The influence of light on hemp development. Tr. VNII konopli, 1935b, 8: 157–164. (R)

Makarevich V.A.: Effects of long days on the progeny of southern hemp: acclimatisation of plants. Kiev: Izd-vo AN USSR, 1953, pp. 115–129. (R)

Makarevich V.A.: Morphological changes and secondary growth in hemp associated with short photoperiod. Tr. VNII lubyanykh kultur. Kiev: 1959, pp. 157–165. (R)

Malutina E.G.: The causes of sex transformation in *Salix triandraf guadrivalis*. Bul. Gl. botan. sada AN SSSR, 1973, 88: 59–61. (R)

Manankov M.K.: The effect of gibberellic acid on fruit formation in grape cultivars that bear functionally female flowers. Fiziol. rastenyi, 1960, 7: 350–354. (R)

Manankov M.K.: The optimal concentration, time and method of introduction in the treatment of grape with gibberellic acid. In: Gibberellins and their effects on plants. M.: Izd-vo AN SSSR, 1963, pp. 226–234. (R)

Manoylov E.O.: A chemical blood reaction as a means to determine the sex in man and animals. Vracheb. gaz., 1923a, 15: 345–347. (R)

Manoylov E.O.: New studies regarding the determination of sex in man, animals, and plants. Vracheb. gaz., 1923b, 21-22: 453–454. (R)

Manoylov E.O.: A chemical reaction that determines the sex in dioecious plants. Tr. po prikl. botanike, genetike i selectsii, 1924, 13: 503–505. (R)

Martin C.: Quelques aspects biochimiques de floraison. Selec. Franç., 1977, 23: 27–31.

Marutyan S.A.: Difference in peroxidase activities in grape seedlings of different sex. Dokl. AN SSSR, 1954, 99: 295–296. (R)

Masters M.F.: Vegetable teratology. L., 1862, p. 138. (R)

Matzke E.B.: Inflorescence patterns and sexual expression in *Begonia semperflorens*. Am. J. Bot. 1938, 25: 465–471.

Maurer G.: Tube Electrophoresis. M.: Mir, 1971, p. 247. (R)

Maurinya Kh.A.: The overall development of different cultivars and hybrids of corn in the Latvian SSR. Agrobiologia, 1956, 5: 16–21. (R)

Maurinya Kh.A.: The link between organogenesis in corn and specific physiological processes. In: Plant Morphogenesis. M.: Izd-vo MGU, 1961, pp. 225–226. (R)

Maurinya Kh.: Techniques to obtain a heterotic progeny in corn. Izv. AN LatvSSR, 1963, 7(192): 91–97. (R)

Maurinya Kh.A., Berzinya-Berzite R.V.: The relation between plant redox processes and sex expression. Nauch. tr. Vsesoyuz. selektsiono-geneticheskovo in-ta, Odessa, 1974, 11: 116–121. (R)

Mazin V.V., Shashkova L.S., Andreev L.N., et al.: The specificity of the influence of kinetin on the formation of amaranthin in *Amaranthus caudatus L.* and on the callus growth in the cotyledons in *Glycine soja L.* Dokl. AN SSSR, 1976, 231: 506–508. (R)

McDavid C.R., Sagar G.R., Marshal C.: The effect of auxin from the shoot on root development in Pisum sativum L. New Phytol., 1972, 71: N 6, 1027–1032.

McMurray A.L., Miller C.H.: Cucumber sex expression modified by 2-chloroethane-phosphonic acid. Science, 1968, 162: 1396–1397.

McPhee H.C.: The influence of environment on sex in hemp, *Cannabis sativa L.* J. Agr. Res., 1924, 28: 1067–1080.

Medvedeva G.B.: On cytology in hemp. Tr. In-ta novykh lubianykh kultur, 1933 (1934), 6: 26–29. (R)

Mekhanik F.Ya.: The treatment of cucumber with acetylene as a means to favor the formation of female, fruit-bearing flowers. Dokl. VASKhNIL, 1958, 11: 20–23. (R)

Mekhti-Zade R.M.: The effects of gibberellin on growth and development of uvae and on other physiological processes in seedless cultivars of grape. In: Gibberellins and Their Effects on Plants. M.: Izd-vo AN SSSR, 1963, pp. 241–244. (R)

Meletti P.: Induzione sperimintale della maschisterilita in Triticum. Nuovogiorn. bot. ital. N. Y., 1961, 68: 299–307.

Mendel G.: Versuche über Pflanzen–Hybriden. Verhandlungen des naturforschenden Vereines. Brünn: Im Verlage des Vereines, 1866.

Mendel G.: Experiments on plant hybrids. M.: Nauka, 1965, p. 159. (R)

Merkis A.I., Putrimas A.D., Marchukaytis A.S.: Interaction between indolylacetic acid and DNA and RNA in plants and its relation to growth. Fiziol. rastenyi, 1971, 18: 78–85. (R)

Merzhanian A.S.: Grape Production. M.: Kolos, 1967, p. 37. (R)

Meyer V.G.: Flower abnormalities. Bot. Rev., 1966, 32: 165–218.

Migal N.D., Zhatov A.I.: Study of processes involved in sex formation in hemp. S.-kh. biologia, 1969, 4: 387–393. (R)

Miginiac E.: Some aspects of regulation of flowering: Role of correlative factors in photoperiodic plants. Bot. Mag. Tokyo, 1978, Spec. Iss., pp. 159–173.

Minenkov A.R.: An attempt at determining the sex. Nauch.-agron. zhurn., 1924, 1: 29–47. (R)

Minina E.G.: Physiological bases in fertilizer introduction techniques. Tr. VIUA, 1935, 8: 75–133. (R)

Minina E.G.: Sex change in plants. Bul. VASKhNIL, 1936, 9: 39–40. (R)

Minina E.G.: The significance of age in determining the sex in plants. Dokl. AN SSSR, 1949, 69: 93–96. (R)

Minina E.G.: Environmental Factors in Sex Transformation in Plants. M.: Izd-vo AN SSSR, 1952, 199. (R)

Minina E.G.: The biological bases for flowering and fruit-bearing in oak. Tr. In-ta lesa AN SSSR, 1954, 17: 5–97. (R)

Minina E.G.: Sex determination in forest woody plants. Tr. In-ta lesa AN SSSR, 1960, 47: 76–161. (R)

Minina E.G.: Sex determination in trees. Avtoref. dis. . . . d-ra biol. nauk. M., 1962, p. 27. (R)

Minina E.G.: The importance of sex transformation in plant selection (the relation between heterosis, polyploidy, and sex expression). Zhurn. obsh. biologii, 1965, 26: 416–427. (R)

Minina E.G.: Rev. of: Djaparidze, L.I. Sexuality in Plants. Tbilisi: Metsiereba. Fiziol. rastenyi, 1966, 13: 377–379. (R)

Minina E.G.: Sex determination and natural growth inhibitors in specific forest conifers. In: Sexual Reproduction in Conifers. Novosibirsk: Nauka, 1973, 1: 146–156. (R)

Minina E.G., Guseva V.A.: The effect of mineral nutrition on sex characteristics in corn. Khimizatsia sots. zemledelia, 1937, 3: 47–60. (R)

Minina E.G., Tylkina L.G.: Physiological studies of the effects of gases on sex expression in plants. Dokl. AN SSSR, 1947, 55: 162–172. (R)

Minina E.G., Kushnirenko S.V.: The role of leaves in plant sex expression. Dokl. AN SSSR, 1949, 64: 261–264. (R)

Minina E.G., Larionova N.A.: Morphogenesis and sex expression in conifers. M.: Nauka, 1979, p. 215. (R)

Mishra R.S., Pradhan B.J.: The effect of (2-chloroethyl) trimethyl ammonium chloride on sex expression in cucumber. Hortic. Sci., 1970, 45: N 1, 29–31.

Mittwoch U.: Sex chromosomes and sex chromatin. Nature, 1969, 204: 1032–1034.

Mittwoch U.: Genetic of sex differentiation. New York, Academic Press, 1973, p. 153.

Mizrahi Y., Richmond A.E.: Abscisic acid in relation to mineral deprivation. Plant Physiol., 1972, 50: N 6, 667–670.

Mizunov G.: Selection of watermelons and melons for early ripening. Kartofel i ovoshi, 1968, 8: 37. (R)

Mohan Ram H.Y., Jaiswal V.S.: Induction of female flowers on male plants of *Cannabis sativa* by 2-chloroethane phosphonic acid. Experientia 1970, 26: 214–216.

Mohan Ram H.Y., Jaiswal V.S.: Induction of male flowers on female plants of *Cannabis sativa* by gibberellin and its inhibition by abscisic acid. Planta, 1972, 105: N 4, S. 263–266.

Molliard M.: Sur la détermination du sexe chez le chanvre. C. r. Acad. Sci., 1897, 125: 792–794.

Molliard M.: De l'hérmaphroditisme chez la Mercuriale et le chanvre. Rev. Gén. Bot., 1898a, 10: 320–326.

Molliard M.: De l'influence de la température sur la détermination du sexe. C. r. Acad. Sci., 1898b, 127: 669–671.

Molotovskyi G.Kh.: On the possibilities in changing the nature of corn. Sov. botanika, 1940, 5/6: 325–331. (R)

Molotkovskyi G.Kh.: Coefficients of polarity and developmental gradation in the formation and distribution of substances in plant organisms. Fiziol. rastenyi, 1956, 3: 243–251. (R)

Molotkovskyi G.Kh.: The appearance of polarity and sex transformation in corn. Dokl. AN SSSR, 1957, 114: 434–437. (R)

Molotkovskyi G.Kh.: The theory of polarity in plant development. Bul. MOIP Otd. biol., 1960, 15: 85–90; part 2, 65–77. (R)

Molotkovskyi G.Kh.: Polarity and sex expression in tissues of corn shoots. In: Proceedings of the XXIst Scientific Meeting of the Chernovitsy State Institute, Section of Biological Sciences. Chernovitsy, 1965, p. 2. (R)

Molotkovskyi G.Kh.: Polarity and heterosis in plants. Vestn. s.-kh. hauki, 1966, 11: 28–34. (R)

Molotkovskyi G.Kh.: Sex, inbreeding, and sterility as a function of the polarity of their development. Vestn. s.-kh. nauki, 1968, 9: 29–35. (R)

Molotkovskyi G.Kh.: Bisexual parity in the development of corn plants. Nauch. tr. Vsesoyuz. selectiono-geneticheskovo in-ta, 1974, 11: 98–104. (R)

Molotkovskyi G.Kh.: Bisexual Parity in Plant Development. Proceedings of the Coordination Meeting on Plant Physiology and Biochemistry, October 14–17, 1975. Kiev: Nauk. dumka, 1975, pp. 38–39. (R)

Molotkovskyi G.Kh.: Bisexual parity in the development and heterosis in plants. Fiziologia i biokhimia kult. rastenyi, 1976, 8: 384–390. (R)

Moniava E.: Comparative Transpiration of Phalaris. Proceedings of the 10th Scientific Conference of the Students of the Tbilisi Scientific Society. Izd-vo Tbil. un-ta, 1948, pp. 14–15. (R)

Moore J.N.: Cytokinin induced sex conversion in male clones of *Vitis* species. J. Am. Soc. Hort. Sci., 1970, 95: 387–393.

Morgan T.H.: Structural Basis of Heredity. M.: GIZ, 1924, p. 310. (R)

Morgan T.H.: Theory of Gene. L.: Seyatel, 1928, p. 312. (R)

Morgan T.H.: The rise of genetics. Sots. rastenievodstvo, 1934, 8: 26. (R)

Morgan T.H.: Development and heredity. M.; L.: Biomedgiz, 1937, p. 285. (R)

Morgan T.H.: The theory of the gene. N. Y., 1928.

Morozov V.A.: Experiments on *Cannabis sativa L.* Tr. Mosk. Lnyanoy opyt. stantsii, 1920, 3: 107–118. (R)

Moshkov B.S.: The role of leaves in the photoperiodic reaction in plants. Sots. rastenievodstvo, 1936, 17: 25–30. (R)

Mothes K.: The role of kinetin in plant regulation. In: Colloq. Intern., Centre Nat. Rech. Sci. P.: CNRS, 1964, 123: 131–140.

Mothes E., Engelbrecht L., Kulajewa O.: Über die Wirkung des kinetins auf Stickstoffverteilungi und Eiweissynthese in isolierten Blatten. Flora (Jena), 1959, 147: N 5, S. 445–447.

Mothes K., Engelbrecht L.: Kinetin-induced directed transport of substances in exised leaves in the dark. Phytochemistry, 1961, 1: 58–61.

Mothes K., Engelbrecht L.: On the activity of a kinetin like root factor. Life Sci., 1963, 11: 852–854.

Moursy H.A., Khalil S.: Effects of gamma radiation and some growth substances on flowering, sex expression and yield of squash. Gartenbauwissenschaft, 1976, 41: N 3, S. 114–118.

Mueller G.C., Gorski J., Aizawa Y.: The role of protein synthesis in early estrogen action. Proc. Nat. Acad. Sci. USA, 1961, 47: N 2, 164–174.

Mukérye S.K.: Contributions to the autecology of *Mercurialis perennis*. J. Ecol., 1936, 24: 38–41.

Muller H.J.: Variation due to change in the individual gene. Am. Natur., 1922, 56: 32–37.

Muller H.J.: Artificial transmutation of the gene. Science, 1927, 66: 84–87.

Mullins M.G.: Hormonal regulation of sex expression in plants. In: Plant growth substance/Ed. F. Skoog. B. New York, Springer-Verlag, 1980, S. 323–330.

Münsche D., Engelbrecht L., Göhler K.D., Conrad K.: Beitrag zur Frage der Beziehungen zwischen struktur und Cytokinin–wirksamkeit. Flora A, 1968, 159: N 3, S. 268–270.

Murakami Y.: Gibberellin like substances in roots of *Oryza sativa*, *Pharbitis nil.* and *Ipomoea batatas* and the state of their synthesis in the plant. Bot. Mag. (Tokyo), 1968, 81: 334–337.

Murakami Y.: The role of gibberellins in the growth of floral organs of *Mirabilis jalapa*. Plant Cell Physiol., 1975, 16: N 2, 337–345.

Muromtsev G.S., Rusanova N.V.: A biological method to determine gibberellin concentration. In: Methods of identifying growth regulators and herbicides. M.: Nauka, 1966, pp. 89–93. (R)

Muromtsev G.S., Agnistikova V.N.: Gibberellins and Crop Yield. M.: Kolos, 1971, p. 127. (R)

Muromtsev G.S., Agnistikova V.N.: Plant Hormones: Gibberellins. M.: Nauka, 1973, p. 270. (R)

Muromtsev G.S., Khrianin V.N.: Antigibberellin effects of chlorocholinechloride. S.-kh. biologia, 1974, 9: 57–60. (R)

Napp-Zinn K.: Modificative Geschlechtsbestimmung bei Spermaphyten. In: Encyclopedia of Plant Physiology/Ed. W. Ruhland. B.; New York, Springer-Verlag, 1967, 18: S. 153–213.

Naugolnykh V.N.: Sexual dimorphism in dioecious plants. Dokl. AN SSSR, 1945, 49: 309–311. (R)

Naugolnykh V.N.: The effect of methylene blue on sex expression in cucumber. Dokl. ASN SSSR, 1948, 59: 995–998. (R)

Naugolnykh V.N.: On the physiology of dioecious plants. Botan. zhurn., 1958, 43: 1562–1571. (R)

Navashin S.G.: New observations in the fertilization of *Fritillaria tenella* and *Lilium martagon*. In: Dnevnik X-ovo syezda russkikh estestvoispytateley i vrachey, 1898, 6: 164. (R)

Naylor F.I.: Effect of length of induction period on floral development in *Xanthium pennsylvanicum*. Bot. Gaz., 1941, 103: 146–154.

Negi S.S., Olmo H.P.: Sex conversion in a male *Vitis vinifera* by a kinin. Science, 1966, 152: 1624–1625.

Negi S.S., Olmo H.P.: Certain embryological and biochemical aspects of cytokinin SD 8339 in converting sex of male *Vitis vinifera* (*sylvestris*). Am. J. Bot., 1972, 59: 851–857.

Neidle E.K.: Nitrogen nutrition in relation to photoperiodism in *Xanthium pennsylvanicum*. Bot. Gaz., 1938, 100: 607–618.

Nelson C.H.: Growth responses of hemp to differential soil and air temperatures. Plant Physiol., 1944, 19: 295–297.

Neskovic M., Konjevic R.: The non-reversible effects of red and far-red light on the content of gibberellin-like substances in pea internodes. J. Exp. Bot., 1974, 25: N 87, 733–739.

Nitsch J.P., Kurtz E.B., Liverman J.L., Went F.W.: The development of sex expression in Cucurbita flowers. Am. J. Bot., 1952, 39: 32–43.

Northmann J., Köller D.: Non transmissible and long-lasting effects of exogenous gibberellin on floral morphology in the eggplant (*Solanum melongena L.*). Planta, 1975, 123: N 2, S. 191–194.

Nozzolillo C.: The effect of gibberellic acid on buckwheat flowers. Canad. J. Bot., 1972, 50: 901–904.

Ochterlony O.: Diffusion in gel methods for immunological analysis. Prog. All., 1958, 5: N 1, 77–79.

Ohkuma K., Lyon J.L., Addicott F.T., Smith O.E.: Abscisic II-an abscission-acceleration substance from young cotton fruit. Science, 1963, 142: 1592–1594.

Orth H.: Die Wirkung das Follikelhormons auf die Entwicklung der Pflanzen. Ztschr. Bot., 1934, 27: S. 565–568.

Osborne D.J., Went F.W.: Climatic factors influencing parthenocarpy and normal fruit-set in tomatoes. Bot. Gaz., 1953, 114: 312–322.

Ostapenko V.I.: The activity of oxidizing enzymes in certain dioecious plants. Botan. zhurn., 1960, 45: 114–116. (R)

Oudin J.: Quelques aspects nouveaux de la spécificité antigénique des protéines du sérum normal. Exposés annu. biochim. méd., 1960, 22: 123–128.

Owens J.N., Pharis R.P.: Initiation and ontogeny of the microsporangiate cone in *Cupressus arizonica* in response to gibberellin. Canad. J. Bot., 1967, 54: N 10, 1260–1272.

Owens J.N., Pharis R.P.: Initiation and development of Western red cedar cones in response to gibberellin induction and under natural conditions. Canad. J. Bot., 1971, 49: N 7, 1165–1175.

Paseshnichenko V.A.: Abscisic acid biosynthesis and metabolism in plants. In: Plant Growth: Primary Mechanisms. M.: Nauka, 1978, pp. 98–126. (R)

Paseshnichenko V.A., Guseva A.R.: The biosynthesis and the physiological roles of phytohormones. Uspekhy biol. khimii, 1974, 14: 254–277. (R)

Pashenko T.E., Redman O.V.: The consequences of shielding cucumber plants with transparent plastic sheets on water regime and other physiological processes. Uchen. zap. LGPI im. Gertsena, 1968, 333: 165–173. (R)

Penel C.: Activité peroxydasique et dévelopement chez *Spinacia oleracea*. Genève, Imprimerie de la section de physique, 1976, p. 160.

Penzig O.: Pflanzen teratologie. 2. Aufl. B., 1921, S. 105.

Peterson C.E., Anhder L.D.: Induction of staminate flowers in gynoecious cucumber with GA_3. Science, 1960, 131: 1673–1674.

Pezard A.: Transformation expérimentale des caractères sexuals secondaires chez les gallinacés. C. r. Acad. Sci., 1915, 160: 260–263.

Pezard A.: Die Bestimmung der Geschlechtsfunktion bei der Hühnern. Ergeb. Physiol., 1928, 27: S. 552–656.

Pharis R.P., Owens J.N.: Hormonal induction of flowering in conifers. Yale Sci. Mag., 1966, 41: N 2, 10–19.

Pharis R.P., Morf W.: Physiology of gibberellin induced flowering in conifers. In: Biochemistry and physiology of plant growth substances/Ed. F. Wightman. Ottawa: Rungepress, 1968, 1341–1356.

Pharis R.P., Morf W.: Sexuality in conifers: effects of photoperiod and gibberellin concentration on the sex gibberellin-induced strobili of Western red cedar (*Thuja plicata Donn.*). Zesz. nauk. UMK Biol., 1970, 13: Z. 23, 45–89.

Pharis R.P., Morf W., Owens J.N.: Development of gibberellin-induced ovulate strobilus of Western red cedar: quantitative requirement of long-day short-long day. Canad. J. Bot., 1970, 47: N 3, 415–420.

Pharis R.P., Morf W.: Precocious flowering of coastal and giant redwood with gibberellins A_3, $A_{4/7}$ and A_{13}. BioScience, 1972, 19: N 8, 719–720.

Pharis R.P., Ruddat M.D.E., Glenn J.L., Morf W.: A quantitative requirement for long day in the induction of staminate strobili by gibberellin in the conifer *Cupressus arizonica*. Canad. J. Bot., 1974, 48: N 3, 653–658.

Pharis R.P., Wample R.L., Kamienska A.: Growth, development and sexual differentiation in *Pinus* with emphasis on the role of the plant hormone, gibberellin. In: Proc. Symp. Mang. Lodgepole Pine Ecosystems, 1975, pp. 106–134.

Pharis R.P., Ross S.D.: Gibberellins, their potential uses in forestry. Outlook Agr., 1976, 9: N 2, 82–87.

Phatak S.C., Wittwer S.H., Honma S., Bukovac M.J.: Gibberellin-induced pollen development in a stamenless tomato mutant. Nature (London), 1966, 209: 635–636.

Pike L.M., Peterson C.E.: Gibberellin A_4/A_7 for induction of staminate flowers on the gynoecious cucumber. Euphytica, 1969, 18: blz. 106–109.

Plakida E.K., Gabovich V.I.: Use of Gibberellin in Grape Production. Kiev: Urozhay, 1964, 102. (R)

Pleshakov I.: The effects of heat treatment of cucumber seed on sprouting efficiency, germinating ability and overall productivity. Sb. trud. nauch.-issled. rabot TSKha, 1951, 3: 81–84. (R)

Pokrovskyi B.V.: Sexual hormones. In: Biochemistry of Hormones and Hormonal Regulation. M.: Nauka, 1976, pp. 246–299. (R)

Polevoy V.V.: Physiology and biochemistry of action of auxin and gibberellin. Dis. . . . d-ra biol. nauk. L.: LGU, 1970, p. 420. (R)

Popov V.I.: The "sexual hormone" of Manoylov. Zhurn. dlya usovershenstvovania vrachey, 1926a, 1: 35–42. (R)

Popov V.I.: The significance and the qualitative determination of the Manoylov's blood reaction. Uspekhi biol. khimii, 1926b, 4: 127–139. (R)

Pravdin L.F.: Sexual dimorphism in Scotch pine. Tr. In-ta lesa AN SSSR, 1950, 3: 190–201. (R)

Pritchard E.: Change of sex in hemp. J. Hered., 1916, 7: N 7, 325–329.

Prusakova L.D., Chizhova S.I., Tsukanova L.D.: The effects of chlorocholinechloride on resistance to lodging and on the productivity and quality of seed in winter wheat. Fiziologia rastenyi, 1970, 17: 1094–1101. (R)

Psarev G.M.: The localization of a photoperiodic stimulus in soy. Sov. botanika, 1936, 3: 88. (R)

Pykhtina T.K.: The effects of naphthylacetic acid on the growth, development, and sex expression in cucumber. In: Morphogenesis in vegetables. Novosibirsk: Nauka, 1971, pp. 273–280.(R)

Pyzhenkov V.I.: Studies of sexual types of *Cucumis sativus* and the evolution of sexuality. Avtoref. dis. . . . kand. biol. nauk. L.: VIR, 1968, p. 30. (R)

Rakitin U.V.: A Manual on Ethylene-Induced Acceleration of Ripening of Tomatoes. M.; L.: Izd-vo AN SSSR, 1950, 64. (R)

Rakitin U.V., Rakitin, V.U.: New ethylene-producing defoliants. Fiziologia rastenyi, 1977, 24: 1004–1013. (R)

Randhawa K.S., Singh K.: Studies on the mechanism of sex expression in an andromonoecious muskmelon (*Cucumis melo L.*) and the factors associated with the induction of flowering. J. Res. Punjab Agr. Univ., 1973, 10: N 3, 301–308.

Resende F.: Auxin e sexo em Hyoscyamus. Bol. Soc. Portug. Cienc. Natur. Ser. 4, 1953, 2: 248–251.

Resende F., Viana M.J.: Gibberellin and sex expression. Portug. Acta Biol., 1959, A6: 77–78.

Ribitska Kh., Engelbrecht L., Mikulovich T.P., et al.: Cytokinin-like active endogenous substances in pumpkin cotyledons and the effect of exogenous cytokinins. Fiziol. rastenyi, 1977, 24: 371–379. (R)

Riede W.: Die Abhängigkeit des Geschlechtes von den Aussenbedingungen. Flora, 1922, 15: N 4, 259–272.

Rieger R., Michaelis A.: Genetisches und cytogenetisches wörterbuch. B. New York, Springer-Verlag, 1958, S. 607. (R)

Rieger R., Michaelis A.: Dictionary of Genetics and Cytogenetics. M.: Kolos, 1967, p. 607.

Robinson R.W., Shannon S., Manuel G., Guardia M.D.: Regulation of sex expression in the cucumber. BioScience, 1969, 19: N 2, 141–142.

Rosa J.T.: Sex expression in spinach. Hilgardia, 1925, 1: 259–275.

Rozanova M.A.: Sex in higher plants. In: Theoretical Bases of Plant Breeding. M.; L.: Gosizdat s.-kh. sovkhoznoy i kolkhoznoy lit., 1935, 1: 145–162. (R)
Rozanova M.A.: Apomixis and heterogamy in *Rosoideae* (Rosaceae family). Dokl. AN SSSR, 1948, 59: 977–981. (R)
Rudenko E.: Sex transformation in hemp. Botan. zhurn., 1956, 41: 868–876. (R)
Rudich J., Halevy A.H., Kedar N.: Increase in femaleness of three cucurbits by treatment with ethrel, an ethylene releasing compound. Planta, 1969, 86: N 1, S. 69–76.
Rudich J., Kedar N., Halevy A.H.: Changed sex expression and possibilities for F_1-hybrid seed production in some cucurbits by application of ethrel and alar (B-995). Euphitica, 1970, 19: N 1, blz. 47–53.
Rudich J., Halevy A.H., Kedar N.: Interaction of gibberellin and SADH on growth and sex expression of muskmelon. J. Am. Soc. Hort. Sci., 1972a, 97: 369–372.
Rudich J., Halevy A.H., Kedar N.: Ethylene evolution from cucumber plants as related to sex expression. Plant Physiol., 1972b, 49: 998–999.
Rudich J., Halevy A.H.: Involvement of abscisic acid in the regulation of sex expression in the cucumber. Plant Cell Physiol., 1974, 15: 635–637.
Ruzinov V.Ya.: Manoylov's Reaction in Fungi. M.: Gos. in-t opyt. agronomii, 1927a, p. 54. (R)
Ruzinov V.Ya.: The use of Manoylov's reaction in fungi. Priroda, 1927b, 10: 826–827. (R)
Ryazanskaya K.V.: Physiological characteristics in the development of monoecious and dioecious hemp. Experim. Botanika, 1956, 11: 318–330. (R)
Rylski I.: Effect of night temperature on shape and size of sweet pepper (*Capsicum annicum*). J. Am. Soc. Hort. Sci., 1973, 98: 149–152.
Ryzhkov V.L.: Genetics of Sexuality. Kharkov: Gosmedizdat, 1936, 250. (R)
Sabinin D.A.: The techniques and the timing in introducing fertilizers and the nature of plants. Dokl. AN SSSR, 1934, 1: 136–140. (R)
Sabinin D.A.: Mineral nutrition as a morphogenetic factor. Bul. MOIP. Otd. biol., 1937, 46: 67–80. (R)
Sabinin D.A.: Mineral Nutrition of Plants. M.; L.: Izd-vo AN SSSR, 1940, p. 306. (R)
Sabinin D.A.: The significance of the root system in the vital activities of plants. In: IX Timiryazevskie chtenia. M.: Izd-vo AN SSSR, 1949, p. 47. (R)
Sabinin D.A.: Physiological Bases of Mineral Nutrition of Plants. M.: Izd-vo AN SSSR, 1955, p. 512. (R)
Sabinin D.A.: Physiology of Plant Development. M.: Izd-vo AN SSSR, 1963, p. 196. (R)
Sabinin D.A.: Selected Works on Mineral Nutrition of Plants. M.: Nauka, 1971, pp. 312–316. (R)
Safarova S.A.: Studies of organogenesis in hemp grown in the Moscow region. In: Plant Morphogenesis. M.: Izd-vo MGU, 1961, 2: 121–125. (R)
Safaryan I.M.: Self-fertilization and pollen fertility in the hermaphroditic flower in watermelon. Biol. zhurn. Armenii, 1966, 19: 112–117.
Saito T., Ito H.: Factors responsible for the sex expression of Japanese cucumber L. XI Role of the leaves. J. Jap. Soc. Hort. Sci., 1961, 30: N 2, 137–147.
Saitov M.M.: On dioecism in hemp. Tr. Kazan. SKhI im. M. Gorkovo, 1964, 44: 73–77. (R)
Sankin L.S., Tulupov Yu.K., Vysochin V.G., et al.: The use of 2-chloroethylphosphonic acid in the regulation of sex expression in melon and in the production of hybrid seed. Sib. vestn. s.-kh. nauki, 1978, 5: 95–97. (R)
Satina S., Blakeslee A.F.: Biochemical differences between sexes in green plants. Proc. Nat. Acad. Sci., Boston, 1926, 12: 192–202.

Satina S., Blakeslee A.F.: Further studies on biochemical differences between sexes in plants. Proc. Nat. Acad. Sci., 1927, 13: 115–124.

Satyanarayana P.: Identification of sex in *Cannabis indica*. Lamk. by Botanical Characters. Madras Agric. J., 1934, 2: 3–8.

Schaffner J.H.: The expression of sexual dimorphism in heterosporous sporophytes. Ohio J. Sci., 1918, 18: 101–103.

Schaffner J.H.: Complete reversal of sex in hemp. Science, 1919a, 50: 311–312.

Schaffner J.H.: Dioeciousness in *Thalictrum dasycarpum*. Ohio J. Sci., 1919b, 20: 25–27.

Schaffner J.H.: Influence of environment on sexual expression in hemp. Bot. Gaz., 1921, 71: 197–219.

Schaffner J.H.: Control of sexual state in *Arisaema triphyllum* and *A. dracontium*. Am. J. Bot., 1922, 9: 72–74.

Schaffner J.H.: The influence of relative length of daylight on the reversal of sex in hemp. Ecology, 1923a, 4: 323–327.

Schaffner J.H.: Observations on the sexual state of various plants. Ohio J. Sci., 1923b, 23: 149–152.

Schaffner J.H.: Experiments with various plants to produce change of sex in the individual. Bull. Torrey Bot. Club, 1925a, 52: 35–47.

Schaffner J.H.: Sex determination and sex differentiation in the higher plants. Amer. Natur., 1925b, 59: 115–117.

Schaffner J.H.: The influence of the substratum on the percentage of sex reversal in winter-grown hemp. Ohio J. Sci., 1925c, 25: 172–174.

Schaffner J.H.: Sex and sex determination in the light of observation and experiments on dioecious plants. Amer. Natur., 1927a, 61: 319–322.

Schaffner J.H.: Control of sex reversal in the tassel of Indian corn. Bot. Gaz., 1927b, 84: 440–449.

Schaffner J.H.: Further experiments in repeated rejuvenations in hemp and their bearing on the general problem of sex. Amer. J. Bot., 1928, 15: 77–79.

Schaffner J.H.: The fluctuation curve of sex reversal in staminate hemp plants induced by photoperiodicity. Amer. J. Bot., 1931, 18: 424–427.

Schaffner J.H.: Observations and experiments on sex in plants. Bull. Torrey Bot. Club, 1935, 62: 387–401.

Schmidt A.A., Perevozskaya N.O.: The physiological and biochemical basis for Manoylov's reaction. Vracheb. gaz., 1926, 13: 650–658. (R)

Schopfer W.H.: Recherches sur le dimorphisme sexuel biochimique. C. r. Soc. Phys. et Hist. Natur. Genève, 1928, 45: N 1, 1418–1419.

Sedlovskyi A.I.: NAA–and gibberellin-induced morphophysiological changes in various cultivars of cucumber. Vestn. MGU. Ser. 16. Biologia, 1972, 5: 111–113.

Senchenko G.I., Arinshtein A.I., Timonin M.A.: Hemp. M.: Izd-vo s.-kh. literatury, zhurnalov i plakatov, 1963, p. 463. (R)

Sergeev P.V., Seyfilla R.D., Mayskyi A.I.: Molecular Aspects of Steroid Hormones. M.: Nauka, 1971, p. 221. (R)

Shain S.S.: Directed changes in the heredity of agricultural crops using different light regimes. In: Directed Changes in the Heredity of Agricultural Crops. M.: Izd-vo AN SSSR, 1963, pp. 83–89. (R)

Shifriss O.: Sex control in cucumbers. J. Hered., 1961a, 52: 5–12.

Shifriss O.: Gibberellin as sex regulator in *Ricinus communis*. Science, 1961b, 133: 2061–2062.

Sidorov B.N., Sokolov N.N.: Purely female types of Castor Bean. M.: Izd-vo AN SSSR, 1945, p. 283. (R)
Sidorov B.N., Sokolov N.N.: A female type of *Ricinus communis*. Dokl. AN SSSR, 1947, 52: 497–500. (R)
Sidorskyi A.G.: The biological reserve in increasing plant productivity. Uchen. zap. Gork. ped. in-ta, 1971, 122: 97–104. (R)
Sidorskyi A.G.: The influence of metabolic processes in the leaves and forming flowers on the directionality of sexual differentiation in plants. Fiziologia rastenyi, 1972, 19: 99–105. (R)
Sidorskyi A.G.: Physiologically active compounds that induce changes in the directionality of sexual differentiation. Uspekhi sovrem. biologii, 1978, 85: 111–124. (R)
Sidorskyi A.G., Sidorskaya E.A.: Change in the sex ratio during the development of cucumber. Ontogenez, 1970, 1: 519–525. (R)
Sidorskyi A.G., Sidorskaya E.A.: Functional sexual differentiation in unisexual monoecious plants. Biol. Nauki, 1971, 11: 80–83. (R)
Sidorskyi A.G., Sidorskaya E.A., Aleshina T.I., et al.: Enzymatic activities in the leaves of unisexual plants. Botan. zhurn., 1971a, 56: 422–428. (R)
Sidorskyi A.G., Sidorskaya E.A., Goskova V.A.: The influence of the earth's magnetic field on the directionality of sexual differentiation in unisexual plants. Uchen. zap. Gork. ped. in-ta, 1971b, 122: 105–114. (R)
Sidorskyi A.G., Sidorskaya E.A.: The role of individual biochemical compounds in sex expression in unisexual monoecious plants. Biol. nauki, 1973, 11: 88–91. (R)
Sidorskyi A.G., Sidorskaya E.A.: Biochemical methods in the study of the evolution of sex in plants. In: 3-yi Vsesoyuz. biokhim. syezd. Riga, 1974, 2: 12. (R)
Sidorskyi A.G., Sidorskaya E.A., Mudretsov G.N., et al.: Some questions on sexual differentiation in unisexual monoecious plants. Biol. nauki 1973, 10: 94–99. (R)
Sims W.L., Collins H.B., Gledhill B.L.: Ethrel effects on fruit ripening of peppers. Calif. Agr., 1970, 24: N 2, 4–5.
Sironval C.: Sur la séparation des sexes dans une population de Chanvres cultivés en jours courts de génération en génération. Bull. Soc. Roy. Bot. Belg., 1959, 91: N 2, 255–265.
Sittin D., Itai C., Kende H.: Decreased cytokinin production in the roots as a factor in shoot senescence. Planta, 1967, 73: N 3, 296–299.
Skarzyński B.: An Oestrogenic substance from plant material. Nature, 1933, 131: N 3317, 766–768.
Skazkin F.D., Aleksandrova A.Z., Repina N.N.: The influence of excessive moistening of soil on the formation of reproductive organs and overall productivity in aestival cereal crops. In: On the Water Supply of Plants Under Diminished and Excessive Water Content of Soil. L.: 1968, pp. 5–11. (R)
Skene K.G.M.: Gibberellin-like substances in root exudate of *Vitis vinifera L.* Planta, 1967, 74: N 3, S. 250–253.
Skene K.G.M.: Increases in the levels of cytokinins in bleeding sap of *Vitis vinifera L.* after CCC treatment. Science, 1968, 159: N 3822, 1477–1478.
Skene K.G.M.: Cytokinin production by roots as a factor in the control of plant growth. In: Development and functions of roots/Eds. Torrey J.G., Clarkson D.T.L.: New York, Academic Press, 1975, pp. 365–369.
Skene K.G.M., Kerridge G.H.: Effect of root temperature on cytokinin activity in root exudate of *Vitis vinifera L.* Plant Physiol., 1967, 42: 1131–1133.
Skoog F., Miller C.: Chemical regulation of growth and organ formation in plant tissues cultured in vitro. Symp. Soc. Exp. Biol., 1957, 11:, p. 18.

Skoog F., Hamzi H.Q., Szweykowska A.M.: Cytokinins: structure activity relationships. Phytochemistry, 1967, 6: 1169–1192.

Skoog F., Leonard N.J.: Sources and structure: activity relationships of cytokinins. In: Biochemistry and physiology of plant growth substances. Ottawa, 1968, p. 1.

Sladky Z.: The effect of growth substances on the differentiation of floral shoots of maize. Acta Univ. Carol. Biol., 1966, N 1/2, Suppl., s. 131–132.

Sladky Z.: Role of growth regulators in differentiation processes of maize (*Zea mays L.*) organs. Biol. Plant, 1969, 11: 208–215.

Sladky Z.: Experimentàlni studium kvetni morfogeneze. Fac. Sci. Nat. Univ. Purkynianae Brunensis Folia. Biologia, 1971, 31: N 8, pp 3–97.

Sladky Z.: Experimental study of floral morphogenesis. III. Study of developmental possibilities of leaf and floral primordia and the origin of fruit in Juglans regia L. Preslia, 1974, 46: N 3, 193–197.

Slonov L.Kh.: Leaf pigments in hemp (*Cannabis sativa*) plants of different sexes. Vopr. botaniki, 1974a, 1: 165–177. (R)

Slonov L.Kh.: Changes in nucleic acid content in the leaves of male and female hemp plants as a function of nutrition and humidity. Fiziol. rastenyi, 1974b, 21: 864–867. (R)

Smith O.: Relation of temperature to anthesis and blossom drop of the tomato together with a histological study of the pistil. J. Agr. Res., 1932, 44: 183–190.

Sogur L.N. (Konopskaya), Gamburg K.Z.: The effect of removing the root system or the epicotyl on cytokinin content in pea seedlings. Fiziol. rastenyi, 1979, 26: 632–634. (R)

Soroka V.P., Zhatov A.I.: Characteristics in the development of the endosperm in monoecious and dioecious hemp. Biol. nauki, 1970, 6: 61–62. (R)

Sotta B.: Interaction du photopèriodisme et des effects de la zéatine, du saccharose et de l'eau dans la floraison du *Chenopodium polyspermum*. Physiol. Plant., 1978, 43: N 2, 337–342.

Splittstoesser W.E.: Effect of 2-chloroethylphosphonic acid and gibberellic acid on the sex expression and growth of pumpkins. Physiol. Plant., 1970, 23: N 4, 762–768.

Sprecher A.: Recherches sur la variabilité des sexes chez *Cannabis sativa L.* et *Rumex acetosa L.* Ann. Sci. Nat. Bot., 1913, 17: 255–257.

Starova N.: Do not trim the crowns of poplar trees. Nauka i zhizn, 1969, 9: 144–145. (R)

Staroseltseva L.K.: Pancreatic hormones. In: Biochemistry of Hormones and Hormonal Regulation. M.: Nauka, 1976, pp. 93–126. (R)

Steinach E.: Geschlechtstrieb und echte sekundäre Geschlechtsmerkmale als Folge der innersekretorischen Funktion der Keimdrüsen. Zentr.-Bl. Physiol., 1910, 24: 78–81.

Steinach E.: Feminierung von Männchen und Masculierung von Weibchen. Zentr.-Bl. Physiol., 1913, 27: N 14, 717–723.

Stoddart J.L., Lang A.: Effects of daylength on gibberellin synthesis in red clover. In: VI Intern. Conf. on plant growth substances: Book of abstracts. Ottawa, 1967, p. 112.

Strasburger E.: Versuche mit dioischen in Rücksicht auf Geschlechtsverteilung. Biol. Zentr.-Bl., 1900, 20: S. 657, 689, 721, 753.

Strasburger E.: Über geschlechtbestimmende Ursachen. Jb. Wiss. Bot., 1910, 48: 427–429.

Strunnikov V.A.: Mechanized Harvest of Cocoons of the Mulberry Silkworm. Tashkent, 1962, 52. (R)

Sukachev V.N.: Dendrology and Introduction to Forest Geobotany. L.: Goslestekhizdat, 1938, p. 295. (R)

Surikov I.M.: The genetics of incompatibility in flowering plants. Gnetika, 1965, 2: 158–164. (R)

Sustikova E., Ginterová A.: Sledovanie uplyvu etrelu na kvitnutie a úrodu uhoriek. Polnohospodárstvo, 1973, 19: N 10, 843–849.

Tadocoro T.: Sex differences from the standpoint of biochemistry. J. Fac. Sci. Hokkaido Imp. Univ. Ser. III, 1930/1933. 1 – 1930; 2 – 1933.

Takasi S., Hideo I.: Factors responsible for the sex expression of the cucumber plant. XIV, Auxin and gibberellin content in the stem apex and the sex pattern of flowers. Tohoku J. Agr. Res., 1964, 14: 227–239.

Takegami T., Yoshida K.: Isolation and purification of cytokinin binding protein from tobacco leaves by affinity column chromatography. Biochem. Biophys. Res. Communs, 1975, 67: N 2, 782–789.

Takhtadjian A.L.: The evolution of placentation in higher plants. Izv. Arm. fil. AN SSSR, 1941, 8: 47–64. (R)

Takhtadjian A.L.: On the Evolutionary Morphology of Plants. L.: Izd-vo LGU, 1964, p. 214. (R)

Tarakanov G.I., Agapova S.A.: The effect of Ethrel treatment on sex expression in cucumber and its possible use in hybrid seed production. Dokl. TSKhA, 1973, 195: 157–162. (R)

Tekhanovich G.A.: The effects of physiologically active substances on sex expression in melon. In: Sb. tr. aspirantov i mol. nauch. sotr. VIR. L.: 1970, 17: 339–346. (R)

Teophrastus: Studies on Plants. M.: Izd-vo AN SSSR, 1951, p. 591. (R)

Thompson A.E.: Methods of producing first generation hybrid seed in spinach. Mem. Cornwell Univ. Agr. Exp. Stn., 1955, 336: 1-48.

Tibeau M.E.: Time factor in utilisation of mineral nutrients by hemp. Plant Physiol., 1936, 11: 731–734.

Timofeev-Resovskyi N.V., Vorontsov N.N., Yablokov A.V.: A Short Essay on the Theory of Evolution. M.: Nauka, 1977, 297. (R)

Tjedjens V.A.: Sex ratios in cucumber flowers as affected by different conditions of soil and light. J. Agr. Res., 1928, 36: 721–726.

Tkachenko G.V.: The effect of gibberellin on growth and fruit bearing in grapevine. In Gibberellins and Their Effects on Plants. M.: Izd-vo AN SSSR, 1963, pp. 235–240. (R)

Tolbert N.E.: (2-chloroethyl)-trimethyl-ammonium-chloride and related compounds as plant growth substances. I. Chemical structure and bioassay. J. Biol. Chem., 1960a, 235: 475–479.

Tolbert N.E.: (2-chloroethyl)-trimethyl-ammonium-chloride and related compounds as plant growth substances. II Effect on growth of wheat. Plant Physiol., 1960b, 35: 380–385.

Tompsett P.B.: Studies of growth and flowering in *Picea sitchensis* (Bong.) Carr. 2. Initiation and Development of Male Female and Vegetative Buds. Ann. Bot., 1978, 42: 889–900.

Torrey J.G., Zobel J.G., Zobel R.: Root growth and morphogenesis. In: The Physiology of the Garden Pea/Ed. by J.F. Sutclife and J.S. Pate, L. etc.: New York, Academic Press, 1977, 12: 119–152.

Tournois J.: Anomalies florales du Houblon Japonais et du Chanvre déterminées par des hâtifs. C. r. Acad. Sci., 1911, 153: 1017–1020.

Tournois J.: Influence de la lumière sur la floraison du Houblon Japonais et du Chanvre. C. r. Acad. Sci., 1912, 155: 297–300.

Tournois J.: Etude sur la sexualisation du Houblon. Ann. Sci. Nat. Bot., 1914, 19: 49–51.

Trendelenburg P.: Hormones: Physiology and Pharmacology. M.: Gosmedizdat, 1932, p. 97.

Tronichkova E.: The modifying effect of Ethrel on hybrid cucumber seed production. Proceedings of the International Conference on Plant Growth Substances. Liblitse, 1978, p. 52. (R)
Tschermak E. von: Über künstliche Kreuzung bei *Pisum sativum*. Z. f. d. Landw. Versuchsw. Öster. 3. u. Ber. Dtsch. Bot. Ges., 1900, H. 5: S. 18.
Tschermak E. von: Die Theorie der Kryptomerie und des Kryptohybridismus. I. Beih. Bot. Zetr.-Bl., 1904, 16: S. 11.
Tschermak E. von: Notiz über der Begriff den Kryptomerie. Ztschr. indukt. Abstammungs- und Vererbungslehre, 1914, 11: S. 183.
Turetskaya R.Kh.: Physiology of Root Formation in Cuttings and Plant Growth Stimulators. M.: Izd-vo AN SSSR, 1961, p. 318. (R)
Turetskaya R.Kh., Kefely V.I., Saidova S.A.: The effects of natural and synthetic inhibitors on different types of plant growth. Fiziol. rastenyi, 1969, 16: 825–831. (R)
Uglov P.D.: The effect of nitrogen on the ratio between male and female cucumber flowers in different soil humidity. Uchen. zap. LGPI im. A.I. Gertsena, 1959, 192: 185–192. (R)
Ustinova E.I.: Characteristic features in the formation of generative organs in corn flower clusters. Botan. zhurn, 1956, 41: 864–867. (R)
Valter O.A., Lilienshtern M.F.: On the diagnosis of sex in hemp. Dokl. AN SSSR, 1934a, 1: 515–521. (R)
Valter O.A., Lilienshtern M.F.: On the diagnosis of sex in plants. In: Tr. Laboratorii physiologii i biokhimii rastenyi Akademii nauk. M.; L.: Izd-vo AN SSSR, 1934b, 1: 127–147. (R)
Valter O.A., Lilienshtern M.F., Chizhevskaya Z.A.: Comparative study of photosynthetic energy in male and female hemp plants. Experim. botanika. Ser. IV, 1940, 5: 72–87. (R)
Van Staden J.: An evaluation of techniques used for extracting endogenous cytokinins from plant material. J. South Afr. Bot., 1973, 39: N 3, 261–267.
Vashenko S.F.: Characteristics in the growth, development, and organogenesis in cucumber (*cv. Nerosimye*) as a function of the environmental conditions. Tr. NII ovoshnovo khoz-va, 1959, 2: 70–87. (R)
Vavilov N.I.: The law of homologous series in hereditary variations. Proceedings of the 3rd All-Russian Breeders' Meeting in Saratov, June 4, 1920. Saratov: Gubpolygrafotdel, 1920, p. 16. (R)
Vavilov N.I.: The use of genetics in Socialist agriculture. Proceedings of the All-Union Conference on the Planning of Genetic and Breeding Research for 1933–37. Tr. po prikl. botanike, genetike, i selektsii. Ser. A., 1932, 4: 19–42. (R)
Vendrovskyi V.: The present day knowledge of the evolution of sex. Uspekhi sovrem. biologii, 1933, 2: 12–28. (R)
Vergely E., Barthelmess I., Hoffmann W.: Der Einfluss von Kurz und Langtag und von Chemikalien auf die Geschlechtsausprägung und den Wuchstyp des Hanfes (*Cannabis sativa L.*). Ztschr. Pflanzen, 1967, 57: N 1/2, S. 26–57.
Vince-Prue D.: Photoperiodism in plants. L., 1975, pp. 312–330.
Vladimirova Z.L.: The influence of heat treatment of seed on fruit bearing in cucumber. In: Referaty dokladov TSKhA. M.: TSKhA, 1952, 16: 272–276. (R)
Vlasenko V.S.: The relation between sex expression and nucleic acid content in developing *Cucumis sativus L.* Biol. zhurn. Armenii, 1971, 24: 93–100. (R)
Vlasenko V.S.: On the relation between growth and the type of sex expression in plants. Uchen. zap. Latv. un-ta, 1973, 183: 66–74. (R)
Volodarskyi A.D.: Immunochemical analysis of the antigen structure of plant tissues. In: Biophysical Methods in Plant Physiology. M.: Nauka, 1971, pp. 14–33. (R)

Volodarskyi A.D.: Immunochemical analysis of heterologous biopolymer systems of the cell. In: Methods of Isolation and Characterization of Protein Components of the Photosynthetic Apparatus. Pushino, 1973, pp. 142–157. (R)

Voskrensenskaya N.P.: Photosynthesis and the Spectral Composition of Light. M.: Nauka, 1965, p. 309. (R)

Voskresenskaya N.P.: Principles in the photoregulation of plant metabolism and the regulatory effects of red and blue light. In: Photoregulation of Plant Metabolism and Morphogenesis. M.: Nauka, 1975, 16–36. (R)

Voskresenskaya N.P.: Photoregulatory aspects of plant metabolism. In: XXXVII Timiryazevskie Chtenia. M.: Nauka, 1979, p. 48. (R)

Vuillemin W.: Les anomalies végétales; leur cause biologique. P., 1926, p. 108.

Wareing P.F., Seth A.K.: Aging and senescence in the whole plant. Symp. Soc. Exp. Biol. Cambridge Univ. Press, 1967, 21: 543.

Wareing P.F., El-Antably H.M.M., Good J., Manuel J.: The possible mode of action of abscisin (dormin) in the regulation of plant growth and development. Univ. Rostock Math. Naturwiss. Reine, 1967, N 4/5, S. 667–672.

Weiling J.F.: Über Geschlechtsunterschiede in der Assimilation und Transpiration bei einigen zweihasigen höheren Pflanzen. Jb. Wiss. Bot., 1940, 89: N 2, S. 157–207.

Went F.M.: Wuchsstoff und Wachstum. Rec. trav. bot. neerl., 1928, 25: blz. 1–116.

Westergaard M.: The mechanism of sex determination in dioecious flowering plants. Adv. Genet., 1958, 9: 217–281.

Wettstein F. von: Über Fragen der Geschlechtsbestimmung bei Pflanzen. Naturwissenschaften, 1924a, 12: 761.

Wettstein F. von: Morphologie und Physiologie des Formenwechsels der Moose auf genetischer Grundlage. Ztschr. indukt. Abstammungs- und Verebungslehre, 1924b, 33: S. 1.

Wettstein F. von: Die genetische und entwicklungsphysiologische Bedeutung des Cytoplasmas. Ztschr. indukt. Abstammungs- und Vererbungslehre, 1937, 73: S. 349.

Whitaker T.W.: Sex ratio and sex expression in the cultivated cucurbits. Am. J. Bot., 1931, 18: 359–361.

White F.R.: Plant Tissue Culture. M.: Izd-vo inostr. lit., 1949, p. 158.

Witsch H. von: Die Entwicklung männlicher und weiblicher Inflorescenzen bei *Xanthium* under Einfluss von Tegeslänge und Gibberellin. Planta, 1961, 57: N 5, S. 357–369.

Wittwer S.H., Hillyer I.G.: Chemical induction of male sterility. Science, 1954, 120: 893–894.

Wittwer S.H., Bukovac M.Y.: The effect of gibberellin on economic crops. Econ. Bot., 1958, 12: N 3, 213–255.

Wittwer S.H., Bukovac M.J.: Staminate flower formation on gynoecious cucumbers as influenced by the various gibberellins. Naturwissenschaften, 1962, 49: 305–306.

Woo S.L.C., O'Malley B.M.: Hormone inducible messenger RNA. Life Sci., 1976, 17: 1039–1041.

Yakushkina N.I., Khrianin V.N.: The influence of gibberellins on contents of nitrogen, phosphorus, and potassium in hemp plants. Agrokhimia, 1967, 1: 130–135. (R)

Yamada I.: The sex chromosomes of *Cannabis sativa*. Rep. Kihara Inst. Biol. Res., 1943, 2: 64–68.

Yang Da-ping, Dodson E.O.: The effect of kinetin on the growth of diploid and autotetraploid root tips of rye in vitro. Canad. J. Bot., 1970, 48: N 1, 19–25.

Yuzepchuk S.V.: Komarov's conception of "species," historical development and reflection in "Flora SSSR." In: The Problem of Species in Botany. M.; L.: Izd-vo AN SSSR, 1958, 1: 130–204. (R)

Zakordonets A.M.: The gibberellin-induced formation of parthenocarpic tomato fruits. Izv. AN SSSR Ser. biol., 1961, 1: 26–29. (R)

Zalkow L.H., Burke N.I., Keen G.: The occurrence of 5α-androstane-3β, 16α-, 17α-triol in «Rayless goldenrod» (*Haplopappus beterophyllus* blake). Tetrahedron Lett., 1964, 4: 217–221.

Zattler F., Chrometzka P.: Über den Einfluss von Gibberelliusäure auf Blütenund Doldenbildung beim Hopfen (*Hemulus lupulus L.*). Theor. Appl. Genet., 1968, 38: N 5, 213–218.

Zavadovskyi M.M.: Sexuality in Animals and Sex Transformation: The Mechanics of Sexual Development. M.; L.: GIZ, 1923, p. 131. (R)

Zhukov M.S., Sazhko M.M.: The effect of gibberellin on the growth and development of hemp. In: Planting and Primary Treatment of Hemp. Kharkov: 1961, p. 83. (R)

Zhukov M.S., Sazhko M.M.: The importance of gibberellin in increasing hemp productivity. Vestn. s.-kh. hauki, 1963, 3: 68–71. (R)

Zhukov M.S., Chailakhyan M.Kh., Kochankov V.G., et al.: The effect of gibberellin on the growth, productivity and technical quality of hemp. In: Gibberellins and Their Effects on Plants. M.: Izd-vo AN SSSR, 1963, pp. 261–269. (R)

Zhukovskyi P.M.: Crop Plants and Related Species. L.: Kolos, 1964, 596. (R)

Zhukovskyi P.M.: Botany. M.: Vysh. Shkola, 1967, p. 667. (R)

Zilber L.A.: The Bases of Immunity. M.: Medgiz, 1968, p. 495. (R)

Zilber L.A., Abelev G.I.: Virology and Immunology of Cancer. M.: Medgiz, 1962, p. 458. (R)

Zimmermann W.A.: Untersuchungen über die räumliche und zeitliche Verteilung des Wuchsstoffe bei Räumen. Ztschr. Bot., 1936, 30: N 5/6, S. 209–252.

Index